GRASSLAND TECHNIQUES AND MANAGEMENT

Related Books

Titles	*Authors*
• Advancements in Micronutrient Research	*A. Hemantaranjan*
• Climate Agriculture and Man Shrinking Biodiversity and Sustainability	*V.K. Sehgal*
• Coactions and Competition in Higher Plants	*B. Politycka*
• Concise Concepts of Nanoscience and Nanomaterials	*N. Kumar*
• Crop Management	*U.S. Walia*
• Crop Production in Salt Affected Soils	*I.C. Gupta*
• Diseases of Horticultural Crops: Nematode Problems and their Management	*P.P. Reddy*
• Dryland Resources and Technology	*L.L. Somani*
• Farm Mechanization for Production	*Dhirendra Khare*
• Fertilizer Manual	*IFDC*
• Fundamentals of Agricultural Statistics	*K.P. Dhamu*
• Fundamentals of Organic Farming and Gardening: An Instructors Guide	*A. McGregor*
• Glimpse on General Agriculture	*P. Laxman Rao*
• Identification of Crop and Weed Seeds	*A.F. Musil*
• Introduction to Crops of India *2nd Ed*	*N.R. Das*
• Kharpatvar Prabandhan (Weed Management)	*J.P. Tiwari*
• Krishi Jagat: JET Evam Krishi Paryavekshak Pariksha Upoagi	*M.L. Jat*
• Manual on Fundamentals of Agronomy	*L.K. Jain*
• Masala Faslo Ki Vignayki Kheti (Hindi)	*Kamini Kumari*
• Moulds and Mycotoxins of Paddy: Incidence, Impact and Management	*M. Surekha*
• Mycotoxins Problem and its Management	*S.M. Reddy*
• Nanotechnology	*J.C. Tarafdar*
• Quinoa *(Chenopodium Quinoa Willd)*	*Dharm Singh*
• Salinity Tolerance in Plants: Methods, Mechanisms and Management 2nd Ed	*B.K. Garg*
• The Use of Saline Waters for Crop Production	*FAO*
• Theory and Practice of Weed Control	*R.J. Stephens*
• Tillage and Crop Production *2nd Ed*	*N.R. Das*
• Training Manual for Organic Agriculture	*Ilka Gomez*
• Vegetable Crops at a Glance	*S.K. Tyagi*
• Wastewater Treatment and Use in Agriculture	*FAO*
• Weeds Identification and Medicinal Use	*U.S. Walia*

GRASSLAND TECHNIQUES AND MANAGEMENT

L. N. Singh
Himachal Pradesh Krishi Vishvavidyalaya,
Regional Research Station,
Dhaulakuan, H.P.

Antaryami Singh
Defence Laboratory (D.R.D.O.)
Jodhpur, Rajasthan

and

Janardan Singh
Himachal Pradesh Krishi Vishvavidyalaya
Palampur (H.P.)

SCIENTIFIC PUBLISHERS

Published by:
Scientific Publishers
5A, New Pali Road, P.O. Box 91
Jodhpur 342 001, India
E-mail: info@scientificpub.com
Website: www.scientificpub.com

Print : 2026

Cover Photograph: A temperate grassland in Western Himalayas.

ISBN: 9788172332280 (Hardbound)
ISBN: 9789387893221 (E-book)

Printed in India

PREFACE

One fourth of world's land surface is under grassland of one kind or the other. Grassland community develop where there is less rainfall to support forest. Annual precipitation where grasslands develop usually varies from 250 mm to 1000 mm. Environment, plants and animals are the important components of grasslands. These are closely interlinked and affect grassland productivity. Livestock production particularly in tropical region comes from natural rather than planted pastures.

Due to high human population pressure, availability of arable land per head is low particularly in developing countries. Because of this there is increasing pressure for cultivation of food crops. Thus, natural grasslands, meadows, forests, waste and barren land and crop residue are the primary source of fodder. Lack of resources as well as knowledge of grassland improvement and management practices is resulting in improper and under utilization of these resources.

Attempt has been made in this book to present methodology to be adopted for the study of botanical composition, ground cover, range condition, forage variety evaluation, herbage yield, livestock production etc. Techniques of seed inoculation, scarification and pelleting, soil test and carrying capacity have also been described. Improvement and proper management of grassland is necessary for getting optimum livestock production. Efforts have been made to present basic principles involved in grassland management, and to give an outline of different improved practices including reseeding of grasses, introduction of legumes, weed control, fertilizer application, grazing management and soil and water conservation. Information about fodder trees and browse which provide green foliage during lean period has also been presented. A chapter has been devoted on principles and method of establishment of artificial grassland which is possible only on good land and if adequate resources are available.

Growth of forage species is mostly confined to a few months in a year, whereas, fodder is required throughout the year. Write up has been included on techniques of hay and silage preparation to enable conservation of surplus fodder for use during lean period. Animal production depends not only on fodder supply but on its quality also. Important aspects of herbage quality and anti-quality constituents have also been documented. It is hoped that information presented in this book shall be helpful in improving basic principles, potential and approach required to be adopted for the improvement and management of grasslands.

At present, grassland resources in our country are not being exploited to their full potential due to lack of knowledge of grassland husbandry, inadequate attention of policy makers and government institution and lack of resources of farmers. There is need of greater attention for optimum utilization of grassland resources to improve livestock productivity in order to uplift economic condition of farmers particularly in developing countries where pressure on arable land is high.

We wish to express our thanks to Dr. (Mrs) Rita Singh for her invaluable assistance in the preparation of book. We also acknowledge our debt to our family members particularly Ms. Savitri Singh whose support has been very necessary and always helpful.

L.N. Singh
A. Singh
J. Singh

CONTENTS

1

INTRODUCTION

A grassland is defined as a natural land surface covered mainly by plants of grass family. It is estimated that grassland comprises about 24 per cent of the world's vegetation. They extend over to about 55^0 N to 40^0 S (Shantz, 1954). Treeless grasslands predominate the temperate regions and are replaced by savannah with varying proportion of trees and shrubs in the tropics and subtropics.

The principal grassland communities of the world are prairies of North America, the pampas of South America, steppes of North and Central Asia and veldt of South and Central Africa. The occurrence of grasslands is largely controlled by climate and climatically determined grasslands occur in environments where mean annual precipitation varies from 250 mm to 1000 mm and mean annual temperature from 0°C to 26°C (Leith, 1975). These climates are mostly found between deserts and forests or in the rainshadow of major mountain ranges. Further more, grasslands occur in areas where, for some period in the year, soil water availability falls below the requirement for forests, yet sufficient precipitation is received during some period of the year to sustain grasses at the dominant, or atleast, major vegetation component (Lauenroth, 1979).

In region where rainfall is higher, forest and where it is less, deserts develop. The type of community that develops in a region depends not on the years with most favourable conditions but on the years with least favourable conditions. Improper use of natural grasslands have resulted in development of desert in many parts of the world.

The climax grasslands in its true equivalent to prairies, steppes, etc. appears to be absent in India, although grassland is common enough as an enterim stage in the ecological succession and may sometimes be even stable at a preclimax under the influences of repeated burnings and regrowth of grasses and other vegetation (Whyte, 1964).

Above ground, net primary production value for 52 climatically determined grassland sites throughout the world was presented by Lauenroth (1979). Mean annual precipitation varied from 136 mm to 1810 mm and above ground net production rate from 42 g/m^2/yr to 1387 g/m^2/yr.

Seventy five per cent of the total geographical area in United Kingdom is under forages and fodder crops. Their yield potential varies 4-14 t DM/ha in different regions. The average yield being realised is around 5-6 t DM/ha (Wilkins and Down, 1984). Grasslands are a major source of animal protein for human consumption producing around 15% of the total protein available to man (Pimentel *et al.*, 1976).

Sown pastures within the subtropics and tropics (30°N to 30°S) are largely confined to areas with annual precipitation greater than 600 mm ranging from the semi-arid subtropics and monsoonal tropics, with a 4-6 months growing period (Figure 1 and 2) to the humid tropics with a year long growing season (Davis and Skidmore, 1966; Cooper, 1967).

Tropical grass pastures that are well watered and heavily fertilized with nitrogen can have carrying capacities of up to 12 beasts/ha. Extremely high gains in cattle liveweight of 1300-2760 kg/ha/yr have been achieved (Henzell and Mannetje, 1980).

Livestock in the tropics and subtropics function in agricultural systems, crop residues provide the main source of feed, and animal drought and nutrient return are crucial to the success of cropping. Majority of the world's herbivores are with developing countries where production per head is much less than in developed countries (Humphreys, 1991). Most of the world's tropical livestock production comes from natural rather than planted pastures. Temperate pastures in India are mostly found on the hills of Himalayan region in the north and Nilgiris in the south (Figure 3 and 4).

Natural grasslands are utilized under different pastoral systems. Pastoral systems may be defined as those where households are more than 50% dependent upon livestock

पेज 2 के बाद

Fig. 1. Degraded grassland in semi-arid region (Kaul & Gogte, 1993)

Fig. 2. A grassland of mid hills

Fig. 3. Temperate pasture at high altitude in western Himalayas

Fig. 4. Cold arid region with scarce vegetation

products, whilst in agropastoral system families are 10-50% livestock dependent. Pastoral systems may be totally nomadic in which case animals owner have no permanent place or residence and cultivate no crops; the farmers move with herd. Trans-humance, or semi nomadism, implies that stock owners have permanent or semi-permanent places of residence, usually near to land on which family may cultivate crops, but travel with the herd for a long periods away from their settlement area.

Area under permanent pasture and other grazing lands in India is estimated to be about 12 million hectare which comes to about 3.92 per cent of the total geographical area (Singh, 1988). India has 15 per cent of the world livestock population but the total milk production is only 7.8 per cent. This is because of low milk yield per animal. At present there is only 8.1 million hectare of arable land under fodder crops which comes to 4.4 per cent of total cropped area (Singh and Majumdar, 1992). As per one estimate the animal requirement are estimated to be 308.1 Mt of green fodder, 353 Mt of dry fodder and 25.4 Mt of concentrate against the availability figure of only 261, 308.5 and 16.5 Mt of green, dry fodder and concentrate, respectively (Mudgal and Pradhan, 1988).

The basic principle of grassland management and utilization is to convert solar energy into plant material and then conversion of this plant energy into animal product. The efficiency of the process depends upon the soil-plant-animal interaction. Manipulation of these factors in order to increase animal production per unit area is the main objective in grassland management. In order to achieve the management equilibrium, the socio-economic studies of native human communities which influence and manipulate forces at work in the ecosystems for increased productivity and biological efficiency should be kept in view (Paroda and Mal, 1995). An appropriate management would help not only to enhance production, productivity and economic returns, but also to maintain the energy cycle in nature and desirable ecobalance.

2

GRASSLAND

Plant community natural or sown, dominated by grasses, herbaceous legumes and other herbaceous species with even small proportion of shrubs or trees is referred as grassland. Grassland could be of natural (native) or artificial type. Natural grasslands are those which have not been at any time cultivated. Artificial grasslands are referred to those located on land that has been tilled and sown with domesticated pasture plants. Natural grasslands extending over a wide area and composed of native grasses, herbs and shrubs valuable for forage and in sufficient quantity to justify grazing use are referred as rangeland or natural pasture.

It is estimated that about 90 per cent of the livestock population in India subsists on natural grasses that are available in grasslands, meadows, forests, waste, barren and fallow lands. Grasslands are located mostly on low fertile poor land not suited to arable farming. The high grazing pressure and lack of adoption of any scientific management practices is resulting in their degradation which is affecting adversely the herbage production and in turn poor livestock production (Singh, 1995).

Herbage production from natural grassland is affected by factors like climate, land capability, botanical composition, ground cover, growth pattern of herbage species, infestation of weeds and bushes, grazing pressure and use pattern.

A. Livestock Population and Grazing Area

As per F.A.O. (1979) estimate based on weighing for size of various classes of livestock, cattle accounts for 70 per cent, sheep and goats 11 per cent and buffaloes 10 per cent of the world's domestic animal units. The total livestock population in

India is estimated to be about 445 million (Table 1) out of which 44.9 % are cattle, 17.1 % buffaloes and 35.0 % sheep and goats (Anon., 1995). Area under permanent pasture is estimated to be 11.8 million hectares (Table 2). In addition to this, there are 19.7 and 15.0 million hectares of barren-uncultivable and cultivable waste land providing grazing to livestock. Forest and other groves are on area of 68 and 3.7 million hectares which also provide grazing of varying degree to livestock in our country. Because of less than 4 per cent of geographical area under permanent pasture, their poor management and only 4.4 per cent of arable land under fodder crops, the availability of fodder is low.

Table 1. Livestock population (in thousands) 1987

States/U.T.	Cattle	Buffa-loes	Sheep	Goats	Oth-ers	Total Live-stock
1	2	3	4	5	6	7
Andhra Pradesh	12375	8758	6871	4876	786	33667
Arunachal Pradesh	310	12	27	108	354	810
Assam	7279	623	67	4134	655	10758
Bihar	20839	4872	1520	15032	1186	43449
Goa, Daman & Diu	112	40	—	17	170	256
Gujarat	6240	4502	1557	3585	168	16135
Haryana	2198	3827	889	674	578	8166
Himachal Pradesh	2244	794	1112	1119	68	5337
Jammu & Kashmir	2765	565	2493	1396	162	7381
Karnataka	10174	4035	4727	3888	357	23181
Kerala	3408	327	30	1581	137	5483
Madhya Pradesh	28549	6365	835	7750	777	44276
Maharashtra	16979	4753	2872	9191	444	34239
Manipur	770	140	16	44	403	1373
Meghalaya	587	28	15	194	285	1009
Mizoram	50	5	1	20	85	160
Nagaland	203	15	1	72	326	617
Orissa	13636	1506	1838	4804	590	2237
Punjab	2830	5575	507	537	224	9673
Rajasthan	10920	6343	9933	12579	275	40913
Sikkim	184	2	11	98	38	330
Tamil Nadu	9342	3129	5881	5921	719	24995

Tripura	827	17	3	441	89	1377
Uttar Pradesh	26320	18240	2180	11320	3087	61147
West Bengal	20311	1163	2312	22694	802	47282
Union Territories						
A & N. Islands	46	14	—	44	—	134
Chandigarh	7	20	—	1	—	32
Dadar & N. Haveli	47	3	—	19	—	69
Delhi	53	285	—	22	—	408
Lakshadweep	1	—	—	15	—	16
Pondicherry	89	9	5	31	—	135
All India	199695	75967	45703	110207	13224	445286

Table 2. Areas under main categories of land uses providing grazing ('000 ha) 1990-1991

State	Reporting area	Forest	Un-cultivable land	Permanent Pastures & other grazing land	Land under misc. trees	Cultivable waste area
1	2	3	4	5	6	7
Andhra Pradesh	27440	16268	2096	843	262	780
Arunachal Pradesh	5544	5200	48	—	44	—
Assam	7852	1984	1541	184	247	104
Bihar	17330	2949	1016	126	291	372
Goa	361	105	13	1	1	90
Gujarat	18821	1888	2667	849	4	1920
Haryana	4378	170	97	23	3	21
Himachal Pradesh	3368	1039	184	1136	48	125
Jammu & Kashmir	4505	2747	295	127	73	136
Karnataka	19050	3074	798	1098	317	446
Kerala	3885	1081	58	2	34	95
Madhya Pradesh	44343	14326	2078	2734	100	1579
Maharashtra	30758	5410	1717	1619	180	1029
Manipur	2211	602	1419	*	24	*
Meghalaya	2239	98	142	—	153	493
Mizoram	2102	1303	20	4	3	74
Nagaland	1532	862	0	0	125	99

1	2	3	4	5	6	7
Orissa	15540	6476	499	726	859	597
Punjab	5032	222	83	10	12	35
Rajasthan	34259	2354	2750	1912	22	5567
Sikkim	710	257	173	69	5	—
Tamil Nadu	13019	3155	509	124	234	290
Tripura	1049	606	1	a	39	1
Uttar Pradesh	29793	5162	1035	303	545	1034
West Bengal	8846	1091	187	7	46	106
A & N Island	789	692	2	9	30	4
Chandigarh	11	1	a	0	a	a
Dadra & N. Haveli	48	20	a	1	—	a
Daman & Diu	10	—	2	a	1	2
Delhi	147	2	10	1	1	11
Lakshadweep	3	—	—	—	—	—
Pondicherry	48	—	a	a	a	2
All India	305017	67985	19660	11804	3703	1501

* Included under Miscellaneous tree crops
a Below 500 ha

B. Climate

Six major types of world climate with a seventh group of undifferentiated highlands, have been described as follows (Trewartha, 1954).

A. Tropical humid climate : Constantly hot or warm, killing frost absent, rainfall 750-1000 mm, intense sunlight. The rain season has one or occasionally two seasonal peaks.

B. Dry climate: Annual water loss through evapotranspiration exceeds precipitation; extends from the equator through the tropics into the middle latitudes. The rainfall is less than 380 mm and erratic, low carrying capacity of grassland. (Figure 5).

C. Subtropical climate : 8-12 months over 10°C, found in middle latitudes except in tropical mountains; seasons distinguished as wet and dry since temperature contrast are small.

D. Temperate climate : Mostly beyond 40° latitude; divided into milder oceanic and more severe continental climates. Found at higher altitude in tropics also (Figure 5).

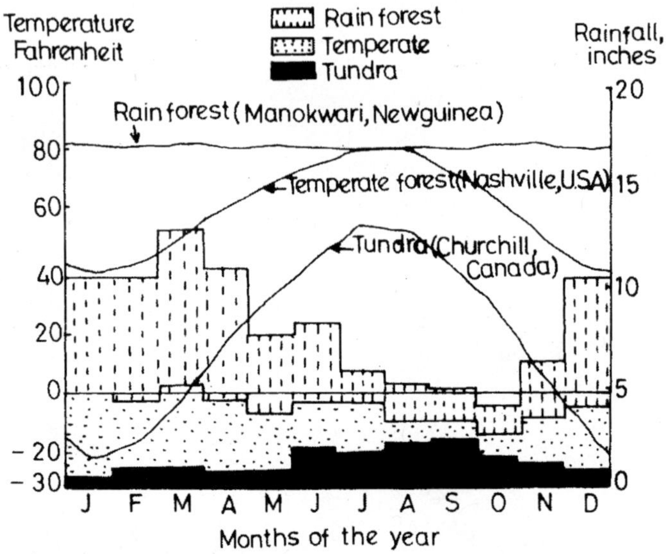

Fig. 5. Average monthly temperature and rainfall at three different sites

E. Boreal (Subarctic) climate : Extending poleward beyond 50° latitude.

F. Polar climate : Found in the higher latitudes of tundra and ice cap.

G. Highlands : Scattered over all continents except Australia, elevation of dominant importance.

Botanical composition, grassland cover, growth pattern of herbage species and productivity of grassland is primarily influenced by climate (Evans *et al.,* 1964; Troll, 1996). Temperature variation and rainfall pattern in particular influence the climatic condition. Average monthly variation in temperature and rainfall of a tropical rain forest at Manokwari, New Guniea, temperate forest at Nashville, Tenessee, U.S.A. and tundra site in Churchill, Canada is shown in Figure 5. Plenty of water is available in a rain forest and temperatures are high and nearly constant throughout the year. Temperature variation is high and growth period is very short in tundra. Temperate site has moderate temperature variation and growing period.

Agroclimatic conditions in India vary from tropical to alpine type depending upon latitude, altitude and proximity to ocean. Soil type, precipitation, physiography, aspect of mountains also influence agroclimatic condition. Twenty different agro ecological regions falling in six ecosystems *viz* arid, semiarid, subhumid, humid, per-humid, coastal and island ecosystems have been identified in our country (Sehgal *et al.,* 1992) as described below.

1. Cold arid eco-region. Found in Ladakh, Gilgit, Lahaul and Spiti districts covering 4.7% area, mean annual temperature of less than 8°C and annual rainfall of less than 150 mm. Length of growing period of less than 90 days.

2. Hot arid ecoregion with desert and saline soil. Found in part of Rajasthan, Gujarat, Haryana and Punjab covering 9.7% of the total geographical area of the country. Mean annual rainfall less than 400 mm with water deficit of 1500-1800 mm in a year. LGP is less than 90 days.

3. Hot arid ecoregion with red and black soils. Found in Anantapur (A.P.) and Bellary, Bisapur Raichur and Tumkur and Chitradurga in Karnataka covering 1.5% of area. Rainfall 400-500 mm. Annual water deficit of 1500-1600 mm. LGP less than 90 days.

4. Hot semi-arid ecoregion with alluvium derived soils. Found in part of Gujarat, Haryana, M.P., Punjab, U.P., Delhi and Rajasthan representing 9.8% of the area. Rainfall 500-1000 mm. The LGP ranges between 90-150 days.

5. Hot semi-arid ecoregion with medium and deep black soils. Found in Gujarat, M.P., Rajasthan and Diu representing 5.4% of the area. Hot and wet summer and dry winters. Annual rainfall 500- 1000 mm. LGP varies from 90 to 150 days.

6. Hot semi-arid ecoregion with shallow and medium black soils. Found in part of A.P., Karnataka and Maharashtra representing 9.5% of the area. Rainfall varies from 600 to 1000 mm. The LGP ranges from 90 to 150 days.

7. Hot semi-arid ecoregion with red and black soils. Found in Andhra Pradesh covering 5.2% of the total geographical area of country. Rainfall 600-1100 mm. The LGP 90 to 150 days.

8. Hot semi-arid ecoregion with red loamy soils. Found in part of A.P., Karnataka and Tamil Nadu representing 5.8% of the total area. Rainfall 600 to 1000 mm. The LGP ranges from 90 to 150 days.

9. Hot subhumid (dry) ecoregion with alluvium derived soils. Found in part of Bihar, U.P., Haryana, Punjab and J&K covering 3.7% of the total area. Rainfall 1000 to 1200 mm. LGP 150 to 180 days.

10. Hot subhumid ecoregion with red and black soils. Found in M.P. and part of Maharashtra covering about 5.8% of the total area. Rainfall varies from 1000 to 1500 mm. The LGP ranges from 150 to 210 days.

11. Hot subhumid ecoregion with red and yellow soils. Found in eastern M.P., South Western Bihar and Mirzapur in U.P. representing 4.3% of the total area. Rainfall 1200-1600 mm. The mean annual temperature of soils more than 22°C. LGP 150 to 180 days.

12. Hot subhumid ecoregion with red and lateritic soils. This constitutes Chhotanagpur Plateau of Bihar, Western parts of West Bengal, eastern Ghats of Orissa and Bastar region of M.P. It represents 8.2% of the total area. Rainfall varies from 1000 to 1600 mm. The LGP varies from 150 to 180 days and at certain places it is 180 to 210 days.

13. Hot subhumid (moist) ecoregion with alluvium derived soils. Found in north Bihar and UP. Occupying 3.4% of the total area. Rainfall 1400-1600 mm. The LGP varies from 180 to 210 days.

14. Warm humid to humid with inclusion of per humid ecoregion with brown forest and podzolic soils. Found in J&K, H.P. and north west U.P. representing 5.6% of the area. Rainfall 1000 to 2000 mm.

15. Hot subhumid to humid ecoregion with alluvium derived soils. It comprises of plains of Brahmputra and Ganga in Assam and Bengal and represents 3.7% of the area. Rainfall varies from 1400 to 2000 mm. LGP in general is more than 210 days.

16. Warm per humid ecoregion with brown and red soils. It encompasses hilly parts of West Bengal, north Assam,

Arunachal Pradesh and Sikkim and represents 2.9% of the total area. Mean annual rainfall exceeds 2000 mm. LGP more than 270 days.

17. Warm perhumid ecoregion with red and lateritic soils. Found in Nagaland, Meghalaya, Manipur, Mizoram and Tripura and represents 3.3% of the total area. Rainfall varies from 2000 to 3000 mm. LGP exceeds 270 days.

18. Hot subhumid to semi arid ecoregion with coastal alluvium derived soils. It comprises the south eastern coastal plain extending from Kanya Kumari to Gangetic delta in West Bengal and represents 2.6% of the total area. Rainfall varies from 900 to 1100 mm. LGP ranges from 90 to 150 days in south to 150 to 210 days in north.

19. Hot humid perhumid ecoregion with red, lateritic and alluvium derived soils. Found in Sahyadris Western coastal plane of Maharashtra, Karnataka States and Nilgiri hills of Tamil Nadu and covers 3.6 % of the total area. The mean annual temperature varies between 25°C and 28°C. The annual rainfall exceeds 2000 mm in most of the area. LGP ranges between 150 and 210 days and at places it exceeds 210 days.

20. Hot humid perhumid island ecoregion with red loamy and sandy soils. It comprises the group of islands of Andaman, Nicobar and Lakshadweep representing 0.3% of the total area. Rainfall 1600 to 3000 mm. The LGP is more than 210 days.

Rainfall pattern combined with temperature and potential evapo-transpiration (PET) determine the length of growing period (LGP). In arid zone, LGP is usually less than 90 days while semiarid corresponds to LGP of 90-150 days. The subhumid regions (moist and dry) correlate with the LGP of 150-210 days, while the humid and per humid regions correlate well with LGP of more than 210 days in a year.

Based on climatic, soil and forage grasses and legumes cover, following five agro-ecological zones have been identified.

A. Arid tract of Rajasthan and adjoining areas.

B. Humid and subhumid tracts of north and north east India.

C. Semi arid tract of south and central India.

D. Sub-tropical and temperate regions in foot hills of Himalayas.

E. Alpine and sub-alpine regions.

C. Grass Cover and Succession

The climax grassland in its true equivalent to the steppes, pampas, prairies in other countries appears to be absent in India, although grassland is common enough as an interim stage in the ecological succession. Grass cover depends upon climate, topography, altitude, soil, soil moisture, grazing, burning etc. Exposure or aspect in montane region also influence floristic composition of grassland. Five distinct grass covers have been recognized in India and are described here (Dabadghao and Shankarnarayan, 1973).

1. *Sehima-Dichanthium* cover

It is found in whole of Peninsular India, central Indian Plateau, Chhota Nagpur Plateau and the Aravalli ranges with approximate coverage of 17,40,000 sq km (Figure 6). Rainfall ranges from 300 mm in Kutch to over 2000 mm or more in Western Ghats. Twenty four perennial grass communities with different ecological statuses occur in this zone. Growth starts after the outbreak of the monsoon in June /July and attains a peak in September. Predominant grasses of this region are *Sehima nervosum, Chrysopogon montanus, Themeda triandra, Eremopogon foveolatus, Heteropogon contortus, Cymbopogon spp., Dichanthium annulatum, Bothriochloa pertusa, Iseilema laxum*, and *Ischaemum aristatum*. The progressive and regressive stages in the plant succession are shown in Figure 7.

Sehima, Dichanthium, Chrysopogon, Iseilema and *Ischaemum* are superior grasses from forage point of view. Legumes associated with this cover are *Atylosia scaraboeoides, Desmodium diffusum, Alysicarpus rugosus* and *Indigofera glandulosa*.

Fig. 6. An over grazed grassland

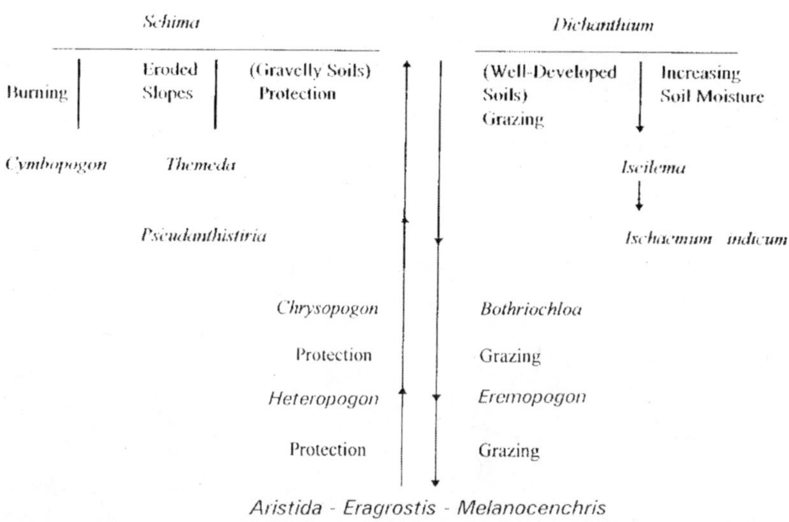

Fig. 7. Succession in *Sehima-Dichanthium* cover

2. *Dichanthium - Cenchrus - Lasiurus* cover

This cover is found in north-west arid and semiarid region of Rajasthan, part of Punjab, U.P. and Gujarat with a potential coverage of more than 4,36,000 sq km and the rainfall coverage of more than 4,36,000 sq. km (Figure 8). The rainfall in the region varies from 100 to 750 mm. The most desirable species of grass in this cover are *Cenchrus ciliaris, C. setigerus, Cynodon dactylon, Dichanthium annulatum* and *Lasiurus sindicus. Cymbopogon jwarancusa, Dactyloctenium sindicum, Desmostachya bipinnata, Eleusine compressa, Eremopogon foveolatus, Sporobolus marginatus, Sehima nervosum, Heteropogon contortus* and *Bothriochloa pertusa* are other important and common grasses of this cover. Important legumes associated with the cover are *Alysicarpus longifolius, Cassia occidentalis, Crotalaria sp., Desmodium triflorum, Rynchosia capitata, R. minima, Tephrosia sp., Sesbania bispinosa* and *Trigonella polycerata*. The succession in this cover is shown in Figure 9.

Fig. 8. A grassland in semi arid region (Kaul & Gogte, 1993)

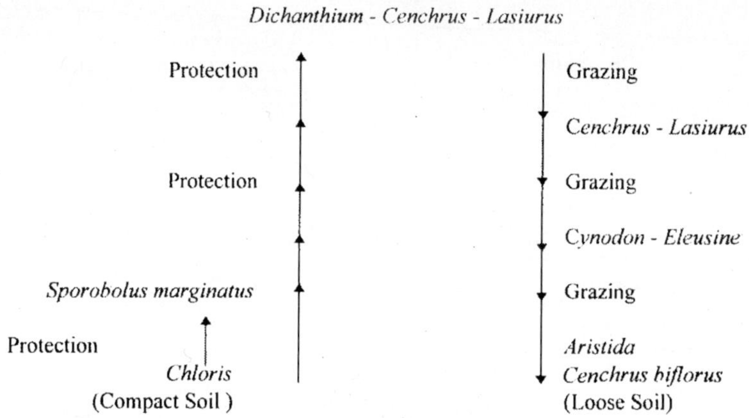

Fig. 9. Succession in *Dichanthium-Cenchrus-Lasiurus* cover

3. *Phragmites - Saccharum* cover

This type of cover is found in an area of about 28,00,000 sq km in humid, subhumid and low lying ill drained areas in north east region comprising Manipur, Assam, Tripura and Meghalaya, and in West Bengal, Bihar, U.P. and Punjab. The cover is characterized by low lying, ill drained lands with high water tables. This cover is dominated by *Desmostachya, Imperata, Saccharum* and *Phragmites genera*. Other important grasses are *Bothriochloa intermedia, B. odorata, Chrysopogon aciculatus, Phragmites karka, Narenga porphyrocoma, Neyraudia reynaudiana, Saccharum arundinaceum, Panicum notatum, Paspalum conjugatum, Sporobolus indicus* and *Vetiveria zizanioides*. Important legumes are *Alysicarpus monilifer, Atylosia scarabaeoides, Cassia tora, Clitoria ternatea, Desmodium triflorum, Medicago denticulata, Pueraria* sp., *Tephrosia candida, Vicia hirsuta.* The progressive and regressive stages of this cover are shown in Figure 10. Most of grasses of this cover are tall and coarse and useful only for thatching purposes. The management of this grasslands for grazing would require drastic alterations in soil conditions to eliminate the coarse water-loving unpalatable grasses (Narayanan and Dabadghao, 1972).

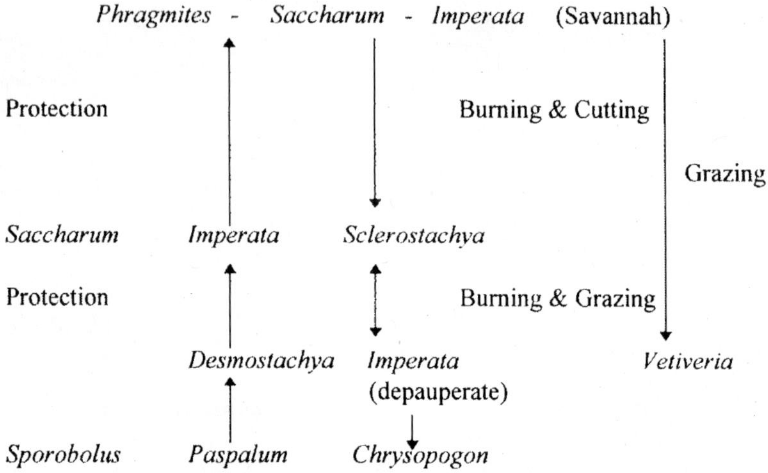

Fig. 10. Succession in *Phragmites-Saccharum-Imperata* cover

4. *Themeda - Arundinella* cover

Themeda - Arundinella cover is found in low and mid hill region varying in altitude from 350 to 2100 m in Himalayas in an area of about 2,30,400 sq km in north east, U.P., West Bengal, Punjab, H.P. and J & K (Figure 11). *Arundinella, Bothrichloa, Chrysopogon, Cymbopogon, Eulaliopsis, Heteropogon, Dimeria, Ischaemum, Themeda, Eragrostis, Eragrostiella, Apluda, Chloris, Capillipedium, Pennisetum* and *Cynodon* are important genera of grasses prevalent in this cover. In *Themeda - Arundinella* cover, *Themeda anathera* represents the highest community and is a dominant species in areas between 650 to 1800 m altitude. On increase in grazing, *Themeda* is replaced by *Arundinella nepalensis*. On further increase in grazing *Chrysopogon fulvus* appears which is replaced by *Heteropogon* and *Bothriochloa* communities. Continued grazing results in presence of *Cynodon dactylon* community. The frequent burning treatment given to grassland at *Arundinella* stage leads to development of *Dimeria* community with *D. fuscescens* as perennial member and *D.*

Fig. 11. Grassland on rocky mountain

ornithopoda as an annual. The progressive and regressive stages in *Themeda - Arunidinella* cover is given in Figure 12.

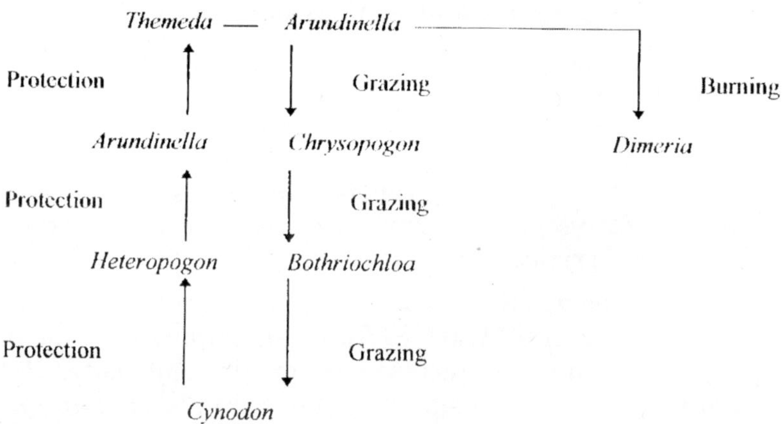

Fig. 12. Succession in *Themeda-Arundinella* cover

5. Temperate and alpine cover

Pastures in temperate and alpine region are at different progressive and regressive stages of ecological succession (Figure 13). It is found in high hills above 2000 m in west and above 1500 m in east (Figure 14). Alpine cover extends up to 5000 m altitude in Himalayas.

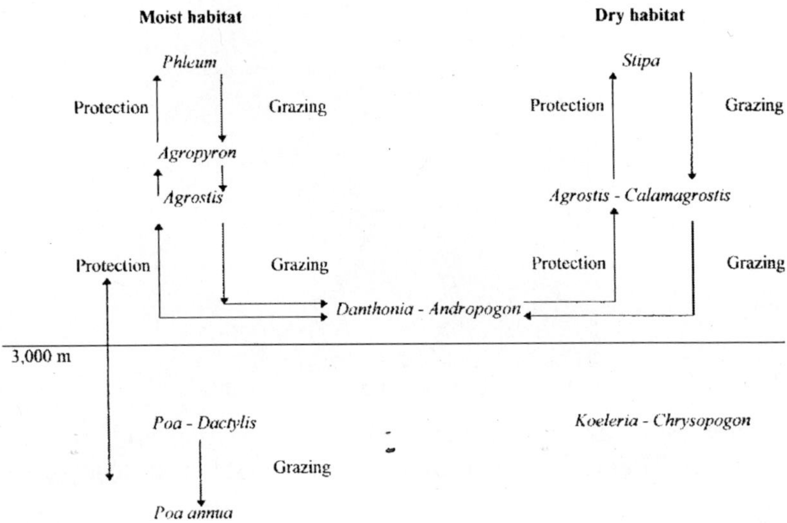

Fig. 13. Succession in Temperate - Alpine cover

Beyond 3000 m altitude, the highest development of grassland consist in establishment of a *Poa - Dactylis* community. On severe grazing *Poa pratensis* is replaced by *Poa annua*. On steeper slopes where drier conditions prevail, development of grassland consists in the establishment of a *Koeleria - Chrysopogon* community with *Koelaria cristata* and *Chrysopogon gryllus* as the principal species.

Grasslands where moisture conditions are more favourable, *Phleum* community with *Phleum alpinum* as predominant species is found. On increase in grazing intensity, *Phleum alpinum* is gradually replaced by *Agropyron canaliculatum* and then by *Agrostis munroana*. On further increase in grazing, they are replaced by *Danthonia* community with *D. jacquemontii* as its chief species. In drier habitats, generally

Fig. 14. A temperate grassland in Himalayas

Stipa community with *Stipa concinna* as a predominant species, is found. On grazing it is replaced by *Agrostis calamagrostis* community which further gets replaced by *Danthonia - Andropogon* community (Dabadghao and Shankarnarayan, 1973). *Bromus, Festuca, Glyceria, Erianthus, Deschampsia, Muhlenbergia, Oryzopsis, Stipa, Trisetum, Deyeuxia, Helictotrichon, Milium* and *Andropogon* are important genera of grasses prevalent in this cover. Species of *Trifolium, Medicago, Melilotus, Lespedeza, Lotus, Lathyrus* and *Indigofera* are important legumes of this cover.

D. Land Capability

Eight broad classes of soil are recognized on the basis of factors which limit their use, such as susceptibility to erosion, natural soil productivity and climatic limitations (Steele, 1951). Classes I-IV describe soils suitable for cultivated crops; class V-VII soil suitable for permanent vegetation as tree crops, forestry and grazing lands; and class VIII rough or steep land, extremely sandy, wet or arid that might have value for wild life or recreation purposes. The soil suitability classification is useful in delineating areas for development of potential grazing lands. Most of natural grasslands, in our country are on land class of V to VII. For establishment of artificial grassland, land class should be of at least class III to IV to enable tillage and reseeding. Different land capability classes are described below (Anon., 1987).

A. Land suitable for cultivation

Class I (Green colour) : Soils in this class have very few or no limitations that restrict their use. Land is nearly levelled and erosion hazard is low. The soils are deep, well drained, and good in fertility and water holding capacity.

Class II (Yellow colour) : Soils have some limitations which reduce the choice of plants. Limitations could be due to slope, susceptibility to erosion, less soil depth, salinity etc.

Class III (Red colour) : Soils have moderate limitations. Limitations are due to moderate slope, soil erosion, overflow, slow permeability, wetness, low moisture holding capacity or moderate salinity.

Class IV (Blue colour) : Soils have severe limitations that restrict the choice of plants. Careful management practices are required and conservation practices are difficult to apply. Limitations of its use are due to high slope, severe erosion, low water holding capacity, excess wetness, overflow, salinity etc.

B. Land not suitable for cultivation

Class V (Dark green or uncoloured) : Soils have little or no erosion hazard. Cultivation is not feasible because of stoniness, overflow, wetness etc. Suitable for pastures, forests etc.

Class VI (Orange colour) : Soils have very severe limitations which cannot be corrected such as steep slope, severe erosion, stoniness, shallow rooting zone, excessive wetness or overflow, low moisture holding capacity, salinity or sodium.

Class VII (Brown colour) : Soils have very severe limitations, that restrict their use largely to grazing or forestation. Difficult to adopt pasture improvement and water control practices.

Class VIII (Purple colour) : Soils have limitations that preclude their use for commercial plant production and restrict their use to recreation, wild life food, aesthetic purposes. Bad lands, rock outcrops, sandy beaches, marshes, deserts, river wash, mine tailings and other barren lands fall in this category.

E. Grassland Production

The productivity of grassland is influenced by a range of biotic and abiotic factors (Figure 15). It is influenced by type of herbage species, ground cover, edaphic factors including the type and texture of soil and its fertility, climatic factors particularly temperature and rainfall and its distribution, and presence of undesirable plant species. The utilization of herbage either directly by animal or under cut and carry system also affect grassland production. The high grazing pressure for longer period leads to elimination of desirable species and eventually decrease in ground cover (Yadav and Singh, 1977).

The factors (Figure 15) which affect grassland production are interlinked to varying degree depending upon situation. Factors depicted away from the centre of dendrogram are generally having lesser influence in comparison to those near the centre.

Grassland production can be judged or assessed on the basis of the production of dry matter, milk or butter, live weight gain or carrying capacity. The net primary productivity (NPP) is defined as the net change in weight of grassland plants between any two points in time, usually over a year. The NPP depends

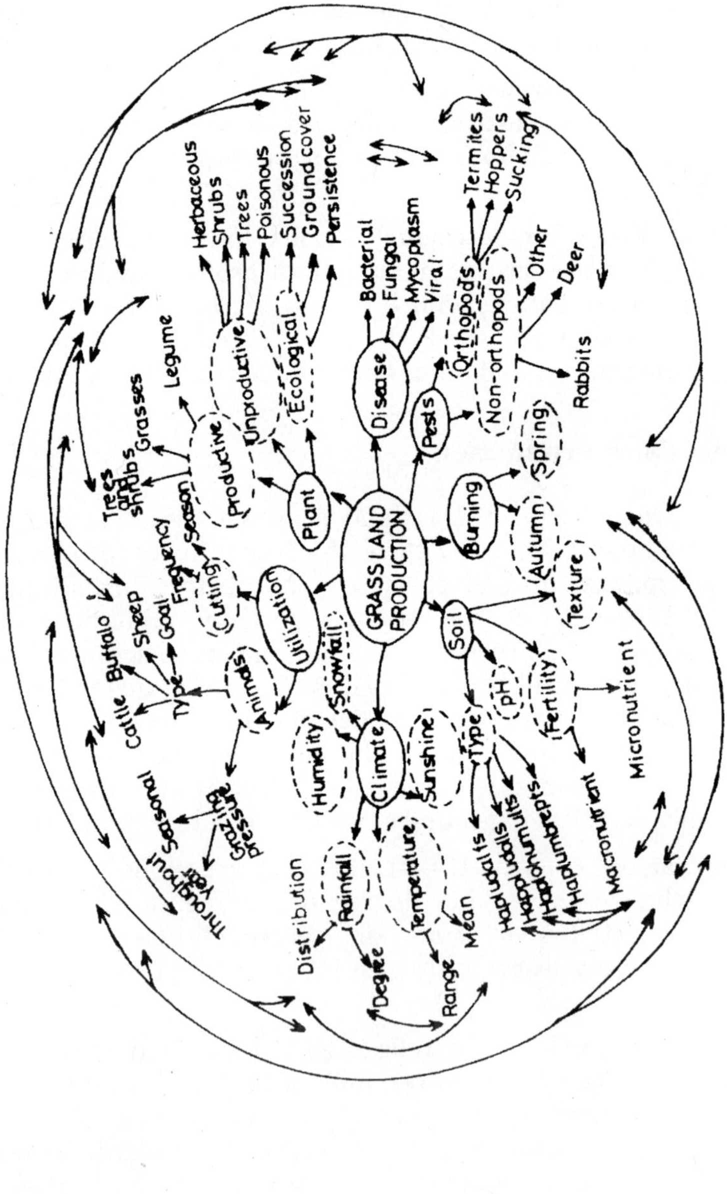

Fig. 15. Dendrogram of grassland production

upon type of grassland climate, soil and management. It has been estimated to be in the range of 10.65 to 19.68 t/ha/yr in 0°-20° latitude and 7.43 to 9.0 t/ha/yr in 20°-50° latitude (Pearson and Ison, 1987). Above ground net production of 0.42 t to 13.77 t/ha/yr has been reported (Christie, 1984). Maximum dry matter yield of tropical grasses may be as high as 85 t/ha/yr with high fertilizer inputs, irrigation and a year long growing season in the humid tropics; the potential is less than half that value in the monsoonal tropics or subtropics (Cooper, 1970). Dry matter production of 6 to 7 t/ha is easily obtained from an improved and well managed grassland. Under ideal condition, dry matter production of 25 t per hectare per year has been obtained (Humphrey, 1969). Growth of tropical grasses ceases at mean daily temperature of 8°C-12°C. A production of 100 kg butter has been reported from a sown irrigated pasture under ideal condition. Under average condition, 600-700 kg butter production per hectare can be obtained.

Dry matter production is also assessed on the basis of carrying capacity or stocking rate. Poor rangeland may have carrying capacity of 20 hectare per animal or less. On little improvement, it could increase to an animal to 2.5 hectare. A improved and well managed grassland may have carrying capacity of an animal to 0.4 hectare.

Grassland productivity can be judged on the basis of live weight gain also. It is low in native pastures particularly in tropical region giving live weight gain of about 50 kg/ha/year. On little improvement these may give 100 to 200 kg liveweight gain per hectare per year. Liveweight gain of 500 to 600 kg/ha is obtained from improved and better managed grassland. Liveweight gain of as high as 1880 kg/ha has been reported from a highly fertilized and well maintained grassland in Australia (Humphrey, 1969). On an improved grassland, liveweight gain of 1.1 kg/day is found under good management condition.

Temperate pastures in general are more productive than tropical. Even with plenty of available feed, cattle may gain only 0.5-0.7 kg/day on tropical pastures compared with 1-12 kg on temperate pasture (Wilson and Minson, 1980). The problem is compounded because the high nitrogen use efficiency of tropical grasses leads to excessive dilution of N and to crude protein levels of less than 6-8% which cause severe reduction in herbage intake. The digestibility of grass, especially stems falls rapidly as the ungrazed material ages. A comparison has been made by Stobbs and Thompson (1975) which shows 70-80 % digestibilty of dry matter in temperate pasture compared to only 50-65 in tropical region.

Milk production per animal is also higher under temperate condition (Table 3).

Table 3. Dry matter digestibility and milk production per lactation from Jersey cows from pastures and concentrate rations (After Driven, 1977)

Diet	Dry matter digestibility (%)	Milk production (kg/head)	Live weight gain kg/day
1. Tropical Pasture			
(a) Immature	60-65	1800-200	0.7-0.9
(b) Semimature	50-55	1000-1400	0.4-0.5
2. Temperate pasture	70-80	3300-3800	0.9-1.2
3. Concentrate ration	80-85	4400-4900	1.2-1.4

Maximum values of plant biomass for various components and net above and below ground production ($g/m^2/yr$) for certain Indian grasslands are given in Table 4 (Malkania and Singh, 1989). There is a wide range in production parameters depending upon agro-climatic conditions, species composition and succession stage. Moisture is a limiting factor in arid and semiarid region, whereas, length of growing season limits production in alpine region. Soil type, its moisture retention capacity, soil fertility, slope, aspect, ground cover, grazing pressure etc. also affect production considerably.

Livestock productivity may be assessed per hectare or per animal. Over all, meat productivity of beef cattle and buffaloes ranges from 20 kg per animal unit per year in developing countries to 96 kg per animal unit per year in developed countries. Likewise milk productivity ranges from 670 to 3100 kg milk per animal unit per year (Henzell, 1983). The productivity of sheep and goats is 40-65 kg meat per animal unit per year.

Table 4. Maximum plant biomass and net production (g/m^2/yr) in Indian grasslands

Parameters	Ecoclimatic zones				
	Semi-arid	Dry sub-humid	Low Hima-laya	Mid Himalaya	High Himalaya
Plant biomass					
Live shoot	76-1408	84-3296	228-706	128-724	217-623
Dead shoot	27-1086	42-1440	36-1239	55-65	103-509
Litter	31-517	30-433	47-232	41-80	43-376
Below ground	86-1145	174-2368	499-1614	577-698	1105-2026
Net Production					
Above ground	38-1950	121-3396	279-1568	219-285	233-3724
Below ground	61-7788	146-1972	358-930	725-897	328-2388

F. Component of Herbage Yield

The above ground herbage of grasses and legumes can be separated into stems, leaves and inflorescences. As stems elongate, side branches develop in grasses and the stem fraction exceeds the leaf fraction of the whole plant. The time at which stem weight exceeds leaf weight varies with species and season.

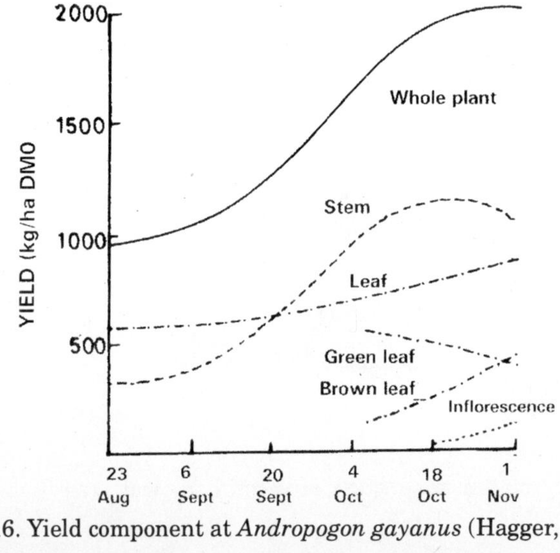

Fig. 16. Yield component at *Andropogon gayanus* (Hagger, 1970)

Hagger (1970) has shown the component of herbage yield of *Andropogon gayanus* which remains leafy until rather late in the season which is desirable attribute (Figure 16). A number of grasses not so suitable for forage becomes quite stalky at early stage of growth.

G. Pattern of Grass Growth

The grass growth in pasture depends upon agro climatic factors and the species. The growth pattern of certain tropical grasses is shown in Figure 17 (Singh, 1990). These grasses show bimodal pattern with major production during summer months (July to September) with some production during spring season (March to April). The temperate species have major production during spring and summer. The typical seasonal pattern of grass growth under grazing type management of a temperate species (*Lolium perenne*) is shown in Figure 18 (Thomas and Young, 1984). Major production occurs in spring when grass growth is very rapid. Growth pattern of temperate grasses under subtemperate condition and

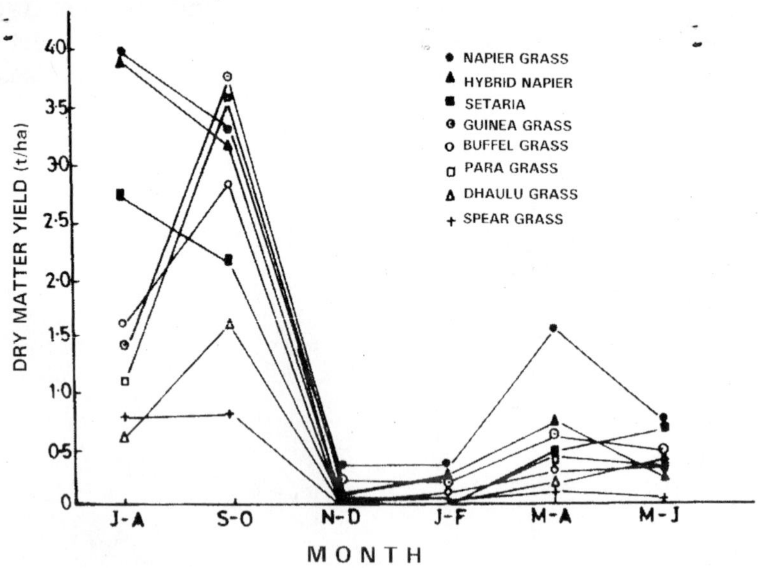

Fig. 17. Dry matter production of certain tropical grasses during different months.

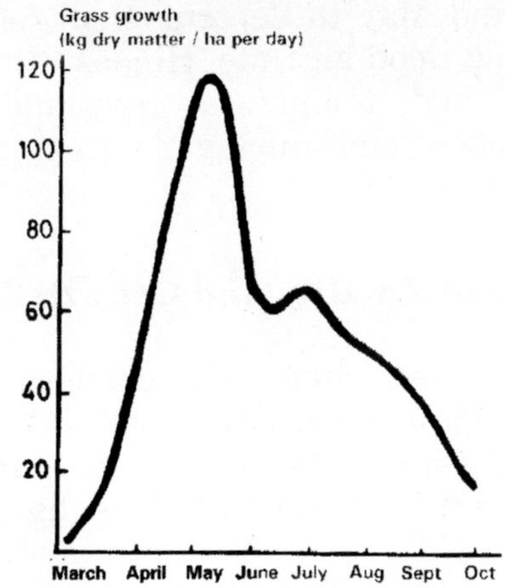

Fig. 18. Seasonal pattern of grass growth (rye grass S 23) under grazing

tropical species under subtropical condition changes depending upon rainfall pattern and temperature variation. Under higher temperature variation, even survival is also influenced.

Seasonal distribution of dry matter and net herbage production in grasslands of Himalayas have been studied by a number of workers (Sharma and Koranne, 1988; Sundriyal *et al.*, 1993) and has been summarized by Sundriyal (1995). The length of growth period in tropical region is mostly during July

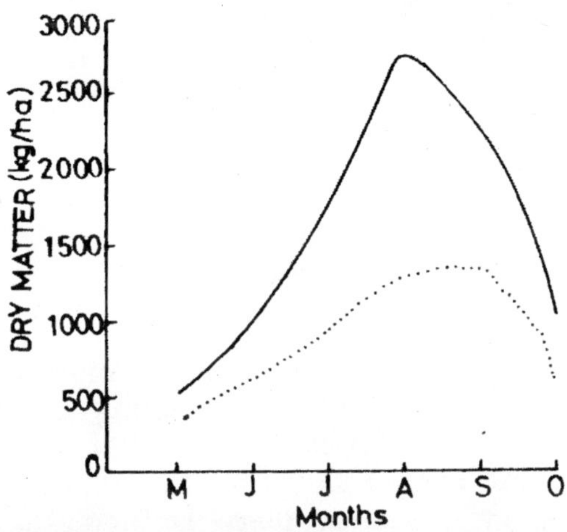

Fig. 19. Seasonal pattern of herbage production in an alpine grassland; entire line and dotted line shows production in ungrazed and continously grazed grassland, respectively

to September and May to September in temperate and alpine region depending upon altitude. Highest dry matter production in subtropical and temperate grassland is found during September-October, and during August in alpine grassland (Figure 19).

H. Herbage Yield, Quality and Growth Stage

As grasses grow, there are changes in water content, digestibility (D value) and crude protein content. The typical pattern of change in these values in a temperate grass (*L. perenne*) is shown in Figure 20. The figure shows a typical

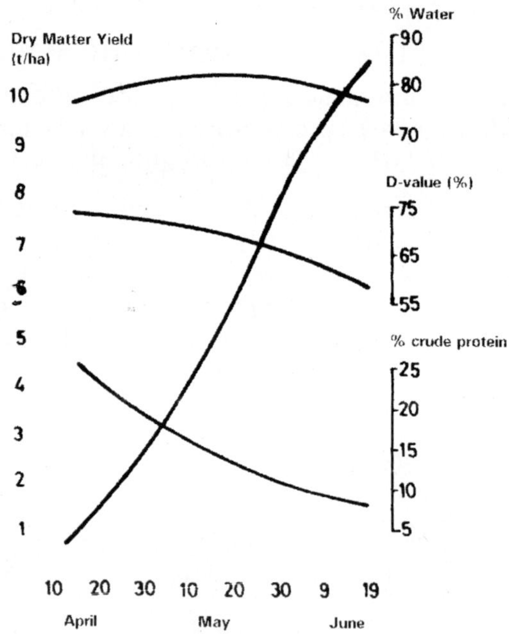

Fig. 20. Changes in yields and quality as grass grows (rye grass S 23)

pattern of change that take place in most of grasses. The values, however, may differ depending upon species and agroclimatic conditions. It can be seen that feed value, in terms of both D value and crude protein content falls as grass is allowed to grow, whereas, there is increase in dry matter production. A balance must be achieved between increasing herbage yield and falling quality.

3

GRASSLAND TECHNIQUES

A. BOTANICAL COMPOSITION

Many techniques have been described and utilized for vegetational analyses. These can be classified into five categories (Brown, 1954) namely :

1. Weight of species component
2. Area covered
3. Number of Individual species or plants
4. Frequency of occurrence and
5. Pattern distribution

1. Weight of species component

It is the most precise and objective method for determining botanical composition of the pasture sward. Each component is expressed as a percentage of the total dry weight of a given sample or treatment. Samples are drawn from herbage cut for yield determination or from samples drawn for botanical composition studies. Cut sample of 0.5 kg from quadrat samples is considered sufficient. Sample is separated species wise and kept separate for dry weight determination and calculation of its percentage contribution to the total dry weight. Height of cutting herbage for botanical analysis should correspond to the height of grazing.

Since hand sorting is a slow process, other methods have also been suggested for measuring species composition. These are based on estimation of relative weight of each species in the standing herbage and estimation of species weight in the cut sample and are useful for rough estimates of changes in botanical composition.

a. Dry weight rank method

It is a method for estimation of botanical composition of grassland on a dryweight basis, without the necessity of cutting and hand separating samples which has been evolved by Mannetje and Haydock (1963). This method is known as dry weight rank method in which in a number of quadrats an observer estimates which species take first, second and third place in terms of dry weight.

The data are tabulated to give proportion of quadrats in which each species receive first, second and third place. These proportions are multiplied by 70.2, 21.1 and 8.7, respectively, added to give the dry weight percentage of each species. The procedure is elucidated in the example given in Table 5. The size of quadrat to be used is not critical, provided that it is large enough to ensure that at least three species will be included in most positions. For subtropical pastures, quadrats of 4, 9, 25 and 40 dm^2 have been used successfully. A quadrat is placed at random on the pasture and the observer records all the species present and estimates which species take first, second and third place in terms of dry weight. Data on 50 to 100 quadrats are recorded and tabulated. This method has been tested by comparing the results with those of hand separating cut samples. No great differences have been found between the two methods.

Table 5. Example of calculations for DW % by dry-weight rank method

Species	No. of quadrats with rank			Proportion of quadrats with rank			Calculations	DW (%)
	1	2	3	1	2	3		
A	70	10	–	.875	.125	–	.875x70.2+.125x21.1	64.1
B	4	25	29	.050	.312	.363	.050x70.2+.312x21.1+0 .363x8.7	13.3
C	6	16	20	.075	.200	.250	.075x70.2+.200x21.1+. 250x8.7	11.7
D	–	26	23.5	–	.325	.294	.325x21.1+.294x 8.7	9.4
E	–	3	7.5	–	.038	.093	.038x21.1+.093x 8.7	1.6
Total	80	80	80	1.0	1.0	1.0		100.1

b. Visual estimation of the composition of cut herbage

A representative cut herbage sample weighing 50 to 500 g depending on species composition of grassland is obtained. This sample is split into 10 roughly equal parts which are spread thinly in separate areas. Two to three observers make independent estimates of the composition of each part, apportioning 10 units among the components. Each observer then has only to total his scores for the whole sub sample in order to arrive at a percentage contribution for each component. Before actual estimation, a few dummy samples are estimated and then separated and the components weighed so that each observer has an opportunity to justify his estimatation (Anon., 1961).

2. Area covered

Cover refers to the proportion of ground covered or occupied by the whole vegetation or individual species. A measure of foliage or canopy cover is usually made for the dense swards of improved pastures and basal cover for swards of rangelands. Many methods and techniques have been used to measure and estimate plant cover. Most commonly used are discussed here.

a. Point Quadrat

A point represents the smallest unit of measure in botanical analysis and as a sampling unit has been referred to as the point quadrat. This technique measures botanical composition and percentage of cover. It is objective and rapid, not dependent on random distribution and allows precise identification of species present.

Points used in vegetation sampling consists of pins for contact of herbage. The apparatus used consists of a frame about 50 cm long made of two horizontally spaced bars about 2.5 cm apart and mounted on legs that could be pushed into the soil. The bars are attached with screws for removal and easy transport. Each bar contains 10 holes 5.0 cm apart so that steel pins could be moved vertically (Figure 21). Frame height is

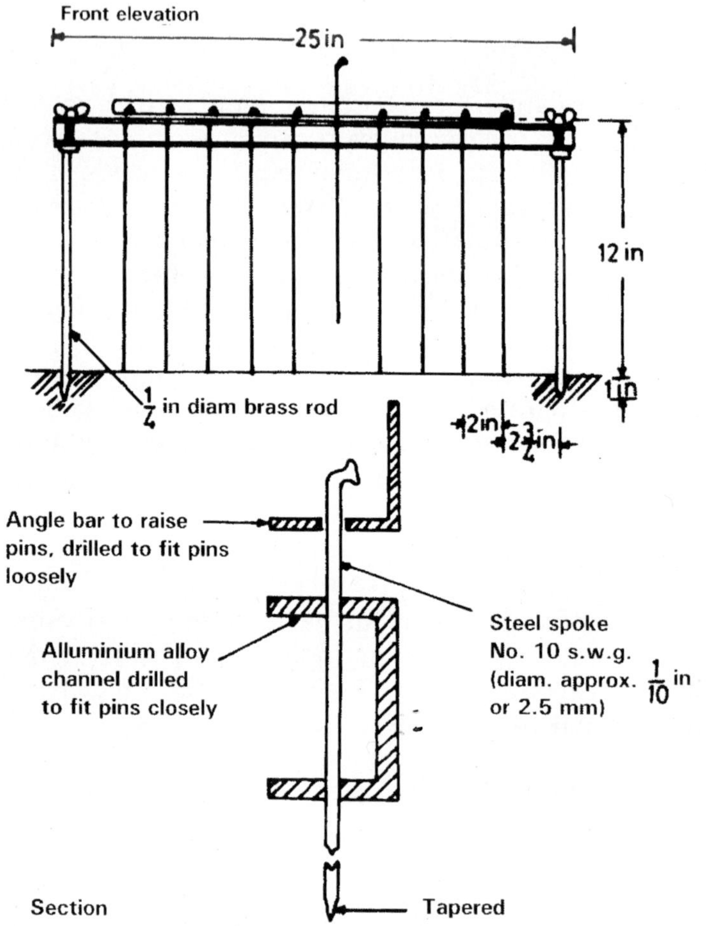

Fig. 21. Frame used in point quadrat method of botanical analysis (Brown, 1954)

about 60 cm but this can be adapted to the vegetation being measured. As the pin moves downward it touches the foliage or bare soil.

The number of samples (hits) to obtain close correlation with weight analysis depends on the heterogeneity of the sward, and whether information is desired on percentage of cover or percentage of species composition. In general, pastures that have a greater number of species or those with sparse vegetation require more points for an adequate sample than those with fewer species and denser swards. In general, 200 to 500 points are sufficient to detect dominant and less frequent species in pasture of 5-50 ha. Data have shown that 300 sites of 5 pins i.e. 1500 observations, gave the same precision as 200 sites and 10 pins (2000 observations). In selecting site or locations for sampling, some investigators distribute the equipment at random. Others layout line transects across the pasture and follow the line, but stopping at random sites.

b. Pace transect or line interception method

Pace transect method or line interception method (Canfield, 1941) can be used to study the floristic composition. This gives an idea of the proportion and relative abundance of the various species present in a plant community. This method consists of determining the base line of the one acre sampling plot and then sampling for composition on five imaginary lines at right angle to base line (Figure 22). The first line is taken at a

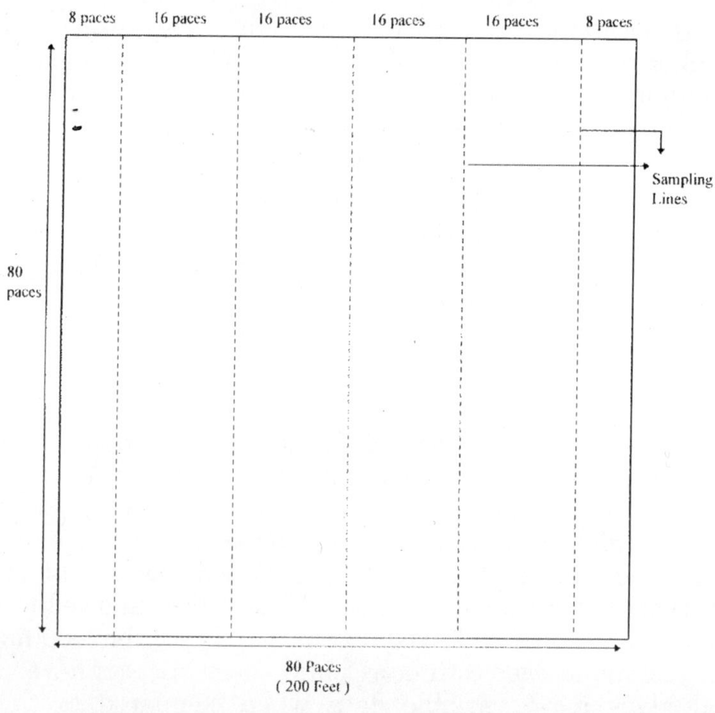

Fig. 22. One acre sampling unit for survey

distance of eight paces (20 ft) from the start up point of the base line and subsequent lines are taken each at 16 paces (40 ft). The total length of the base line is 80 paces.

For sampling the vegetation, the examiner walks along the transect line starting from the base line and data is recorded at every fourth step by putting closely on previously marked line on the toe of his right foot. If a grass plant is present at the toe, point hit was secured and recorded. If effective hits are secured at all sampling points on the five transect lines, a total of 100 hits would be secured. If some hits are missed, additional transect lines are taken. The total hits on each grass species are counted to arrive at the percentage composition.

c. Step point method

Evans and Love (1957) described the step point method for use in rangelands. This technique comprises a combination of following two procedure.

(i) Location of a point at the tip of a shoe after a specified number of paces. The tip of shoe was held at an angle of about 30^0 a pin thrust downward into the sward. A record of the first hit was noted by species or bare ground. This gave a measure of species composition and ground cover.

(ii) A visual estimate of ground cover within a 929 cm^2 quadrat placed on to the sward at the shoe toe after each 10 paces. Total plant cover was measured with greater precision than species composition.

d. Loop frequency

This is a combination of the line interception method and the point quadrat in which vegetation within a 1.9 cm loop is recorded instead of vegetation contacted by a point (Parker and Harris, 1959). In estimating plant cover or species composition, a transect line is laid out and the loop lowered at specified intervals until it contacts foliage (first hit) or allowed to rest on the sward or bare soil (basal coverage). Since the loop has area, the technique over estimates plant cover, detects fewer species and gives more variable data when compared to the point quadrat and line-interception methods.

3. Number of individual species or plants

It is a quantitative measure of population or abundance expressed as density calculated based on number of plants or tillers per unit area or as percentage composition-ratio of one species to the total number of individuals of all species (Brown, 1954). In making counts it should be decided whether to count tillers or whole plants. In a dense sward, individual plants of bunch type, rhizomatous and stoloniferous grasses or trailing legumes can not be detected easily.

This method is useful in studies of percentage germination, establishment, regeneration, persistence of perennial species, disease survey and estimates of seed yield. The number of individuals as a measure of abundance is related to cover within a given growth habit, but should not be used to estimate vegetative cover in general (Crowder and Chheda, 1982).

The counting of number is done by using quadrats or transects to delimit an area, then recording the total number of individuals or tillers. A scale of 1-5 or 1-10 may refer to numbers or percentage.

Stapledon's percentage frequency method

In this method, average percentage of tillers of different species are computed on the basis of their percentages in a number of quadrat samples taken in grassland (Brown, 1954). The method is simple, very convenient and fast. However, it does not reflect herbage composition on weight and is influenced by species to a great extent. The grass species having very thin tillers with low plant height may show high percentage which may not be true reflection of their contribution in herbage yield.

Botanical composition can also be studied by analysing sample of plant fragments collected from oesophagus fistula (Mannetje, 1978). By this method, composition of only grazed pasture can be determined as grazing animals have liking and disliking of species depending upon their palability.

4. Frequency of occurrence

It is expressed as the number of samples in which a given species occurs. Thus, it refers to whether a species is present or absent and not to number of plants (density). A common method for measuring frequency of occurrence is to toss a quadrat randomly onto the sward a prescribed number of times, then to record the presence or absence of a species.

Another method is to walk over the pasture stopping after a predetermined number of paces and noting the presence or absence of a species at the toe of shoe. The entire pasture must be traversed. This method gives no information about the contribution of a species to total herbage production.

5. Pattern distribution

Plant species are not always evenly distributed over a large pasture area. Soil fertility, depth of soil, gradient, topography, exposure (northerly/southerly) drainage etc affect species distribution. Divergent areas should be delineated before sampling. The differences in life cycles and growth patterns of species significantly contribute to variation in botanical composition, ground cover, frequency of occurrence and herbage production. Such factors must be kept in view in sampling time, interval and size for botanical composition studies.

Remote sensing

Remote sensing along with the related technology of Geographic Information System (GIS) and Global Positioning System (GPS) provide new opportunities for intensive analytical examination of rangelands (Tueller, 1995). The perspective offered by remote sensed data gives opportunity to look at large land areas where extensive rather than intensive management is often required. The pixel modelling which is a quantitative assessment of proportion of rock, soil, litter, plant species or species groups inherent in individual pixels eventually lead to the ability to assess forage and browse productivity, suitability for class of grazer or browser, carrying capacity, forage utilization and the monitoring of vegetation and soil surface changes.

Remote sensing provides information for resource interpretation including panchromatic colour and color infrared (CIR) aerial photographs, satellite images and digital data products and aerial videodata, all in various wave lengths and radar data. Photographs are usually in 1:500 to 1:2000 scale and are useful for identifying many species and making detailed measurements. Among spacecraft providing remote sensing data are the Landsat Multispectral Scanner (MSS), the Landsat Thematic Mapper (TM), the SPOT (Systeme Pour & Observation de la Terre) and the AVHRR (Advanced Very High Resolution Radiometer). Orbital characteristic of the space craft result in repeat coverage of the same ground area at regular intervals thus making such data potentially very useful for monitoring.

AVHRR have an approximate scale of 1:10,000,000 and pixels that vary from 1.1 to 8 km depending upon whether the instrument is in Local Area Coverage (LAC) or Global Area Coverage (GAC) made. Microwave systems have potential for rangeland monitoring and include both Side Looking Airborne Radar (SLAR) and Synthetic Aperature Radar (SAR). These are not constrained by either night (darkness) or cloud cover. Thermal infrared Multispectral Scanner (TIMS), an aircraft borne scanner is useful for mapping of soils. High spectral resolution systems (hyperspectral remote sensing) including AIS (Air-borne Imaging Spectrometer), AVIRIS (Airborne Visible and Infrared Imaging Spectrometer) and HIRIS (High Resolution Imaging Spectrometer) are quite sofisticated remote sensing technology. It requires high level of understanding of the spectral characteristics in order to make its use in study of botanical composition, grassland productivity, monitoring etc.

B. GROUND COVER

Ground cover is defined as the area occupied by a vertical projection of each species as it grows. It is expressed, for each species independently, as a fraction of the whole area. No account is taken of the overlapping foliage of the same species. The degree of ground cover occupied by vegetation is referred as plant cover or density.

1. Square foot density method

This method was developed by Stewart and Hutchings (1936). This method consists of taking a circular plot of 100 sq. ft with a circle of 5.64 ft radius (5 feet 7 inches). The unit of measure of density is a square foot of ground completely covered with vegetation when viewed directly from above. This concept of measurement is developed by picking the plants on a plot and placing them with in the square foot area marked on the ground in such a way that they cover the ground fully without overcrowding. Such concept is developed for different species with varying growth forms. Using the square foot as a unit of measure, the examiner views the vegetation enclosed in the 100 Sq ft area and shifts the vegetation mentally into one square foot units. The total number of such units directly gives the percentage plant cover. A well maintained lawn is an example of a 100 percent plant cover, while a bare area represents the opposite extreme.

Depending upon situation, estimates are made generally on all the individual species or on principal species. In some cases, the density estimates can be made in one lot for the entire vegetation. The total number of samples taken for these studies are generally ten, but sometimes because of limitation of time, observations are made on fewer circles.

2. Percentage area estimation

A small area is delimited by a quardrat and examined from directly above. An estimate is made of the proportion of ground cover by each species. The vegetal cover and bare soil should total 100. To facilitate estimation, the quardrat can be subdivided into smaller units and each rated separately (Brown, 1954).

At GRI Hurley, most observers find a 6-inch quadrat a satisfactory unit on which to focus for visual estimation of ground cover. A fresh set of random sites is taken for each successive analysis. From 10-20 quardrats are examined in each plot. Each species is considered in isolation, and the fraction of the quadrat area covered by it recorded in tenths. When a species covers less than 0.1 of the area, it is recorded as a trace. It is considered desirable to restrict the use of ground cover estimates to comparisons between species of similiar growth habit, or to the tracing of changes in single species.

Ground cover is also assessed on visual estimate made on 0-9 scale as per diagram and method given by Camlin and Stewart (1976). High score indicates high persistence. Zero scale indicates no presence and 9, hundred per cent cover of ground by sown or referred species (Figure 23). One to 8 scale shows varying degree of cover. It is a quick and quite reliable method once the examiner has gained experience in scoring. Usually means of scores by three independent observers are taken for analysis.

3. Plant or tiller density

Plant or tiller per unit area is a parameter useful in assessing establishment of species, botanical composition, cover etc. The measurement is made on a very small area in a given time. For counting of seedlings in young leys or plants in tufted vegetation, 6-inch or 12-inch quadrats are normally used. For counting of shoots in dense, heterogeneous vegetation, 3 inch and 6 inch quadrats are used for shorter and taller herbage species. In all these cases, estimates are based on 10-20 quadrats per plot on each occasion. In studies of seedling development or mortality, permanent quadrats or whole microplots of up to 1 sq. yard are mapped at intervals with the aid of a pantograph (Anon., 1961).

C. FORAGE VARIETY EVALUATION

A herbage garden is maintained for preliminary observation of introduced forage species or new varieties. Introduced material is spaced planted in rows 60 cm to 100 cm apart depending upon species with a plant to plant spacing of 30 cm

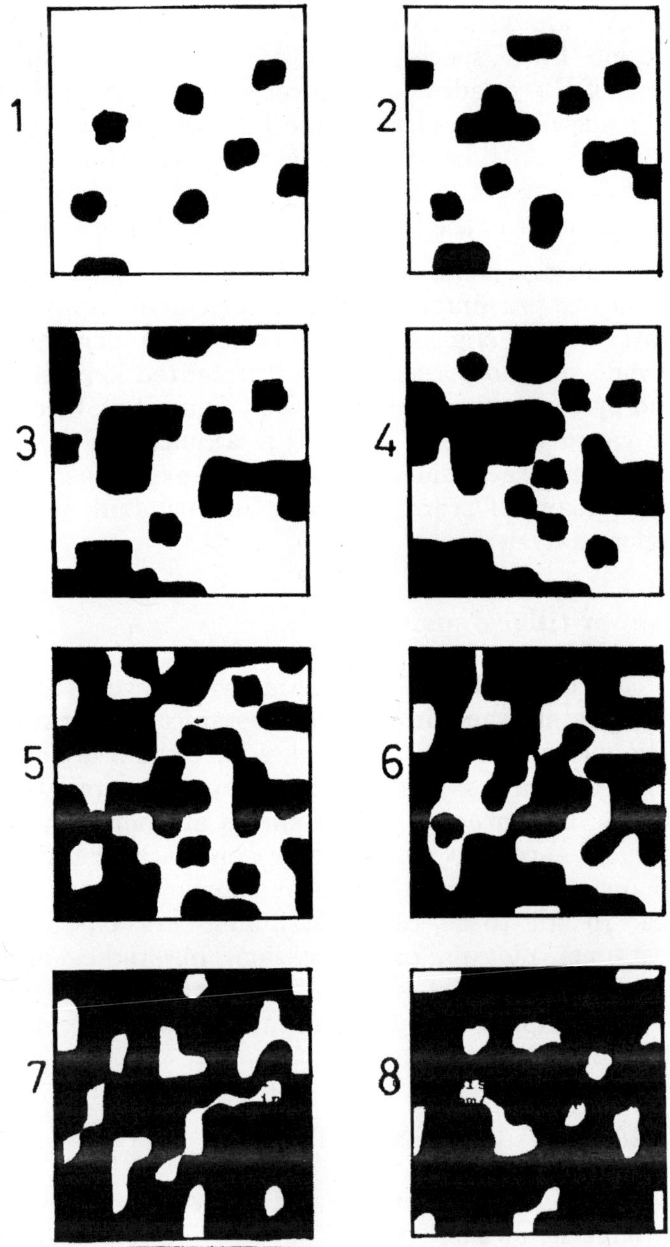

Fig. 23. 1-8 scales of persistence (ground cover) in 0 - 9 scale system (Camlin & Stewart, 1976)

(Figure 24). A plot with 24 plants is usually sufficient for preliminary screening. Observations on establishment, vigour, growth habit, plant height, tiller number, leafiness, onset of flowering, forage yield, regeneration, seed setting, winter hardiness, drought tolerance, disease and insect pest infestation are recorded. The mean date of flowering or ear emergence is recorded as the date on which 50% of plants have reached this stage. Plants in one-half of each plot are cut for herbage yield and regeneration data, while the remainder are left to flower for recording various other observations.

Promising material is advanced to next stage of evaluation under sward condition alongwith standard checks in replicated trial. If evaluation is to be done after manual harvesting, plot size may be kept between 4 to 10 m^2 (Figure 25). Plot size (paddock) in grazing trials should be kept between 130 to 200 m^2 for 4-8 sheep for two to three days and 600 to 700 m^2 for cattle or mixed stock. In grazing treatments per se, it is not always necessary to have large plots. Certain aspects of grazing such as time, frequency and duration and class of livestock have been studied in small plots by employing fences to keep stock out rather than in. Stock are kept in an accommodation paddock with controlled access to each part in a compact array of small plots. At any moment they range both the paddock and any plots which are due for grazing. The paddock has only to be large enough to carry one or two cattle and a few sheep whenever there are plots due for grazing (Anon., 1961). The accessibility of herbage to the animal depends not only on its position but also on its physical characteristics. In estimation of herbage yield by harvesting, samples are cut at a lower level during periods of general scarcity than in periods of plenty and at a lower level (0.6 to 2.0 cm) for sheep pasture than for cattle pasture (2.0 to 5.0 cm).

In grazing trials, cages of 3x1 m are used to exclude grazing from where samples could be collected at different stages of trials for botanical and herbage yield studies. The quantity of herbage grown or consumed can be estimated by difference method, 'in' and 'out' samples based on paired sampling units (Anon., 1961; Anon., 1984). The sites for each pair of units are fixed before cutting takes place. If for instance 4 pairs of units are to be cut in a plot, 4 sites are taken at random, and 4 more either matched with them visually or sited as close to them as possible. One unit of each pair is then allocated at random to the 'in' sample and one to the 'out' sample. This method is used where plots are grazed intensively, i.e. for short periods at intervals of at least 3 or 4 weeks. For comparison of swards or treatments, the number of pairs of sampling units cut in the course of single grazing rotation is usually about 16 per treatment (all replicates included) if the area of each unit is between 1 and 2 m^2.

New varieties of grasses and legumes are developed for specific attributes such as seasonal growth rythm, persistency, resistance to drought, winter hardiness, disease and insect pest resistance, specific adaptability, vigour, regeneration, stability, nutritive value, fertilizer responsive, palatability, grazing, conservation, digestibility, seed production, dry matter yield etc. Whilst all these characteristics are important, their relative importance will vary in accordance with the demands of different livestock systems and environments in which they are grown. Newly developed varieties are aimed to have at least 10% superiority in yield or have advantage of desirable attributes.

The varietal evaluation of forages is an essential step in development of new varieties. The evaluation should be carried out at a number of locations to assess genotype-environment interaction and the stability of performance. Since forages are grown not as an end products in themselves but as a raw material for animal products, simple evaluation on dry matter yield may be misleading. Thus, forage varieties should be evaluated for :

1. Chemical composition with respect to nutritive value
2. Antiquality component if any
3. Production in terms of different forms of animal product
4. A comprehensive range of practical grassland systems
5. Different levels of intensity of utilization
6. Performance at different locations
7. Period of time that reflects their intended use in practice
8. Grazing or conservation
9. Association with other companion grasses and legumes

Newly bred herbage variety should be evaluated through different stages. A scheme of forage varieties evaluation in vogue in United Kingdom is given below (Walters, 1980).

Phase - I *Varietal testing in small plots*

All new varieties are tested on small plots ($8m^2$) in comparision to control and data on establishment, vigour, seasonal growth, rythm, persistency, hardiness, annual dry matter yield, regeneration, reaction to disease, insect pests, tolerance to drought, digestibility, chemical composition is collected. At the end balance sheet of the full range of characteristics is prepared and if a variety qualifies, it is advanced to next phase of evaluation.

Phase - II *Farm type management trial*

Variety is sown as a single variety sward and evaluated. This is to assess to what extent which relative advantages found in earlier stage of testing are maintained under more practical farm type management. The range of management applied may include different dates and frequencies of cutting for conservation or grazing, different cutting heights, or grazing pressures, different rates and times of fertilizer application. Potential varieties are advanced to Phase III.

Phase III *Seed mixture and feed value assessment and farm trial*

Forage varieties entering this phase of evaluation are tested in mixture with other grass or legume varieties in different proportion in small plot replicated field trials and compared with standard seed mixtures. Detail observation on their performance is recorded. Simultaneously, feed value assessment of varieties in terms of voluntary intake of digestible organic matter (DOMI) is made in indoor comparative feeding trials using sheep as tester animals. Grazing and feeding trials are carried out for milk, butter or meat production. Potential varieties are then grown on selected commercial farms representing different grassland systems under different agroclimatic conditions. The farmer is given questionnaires for recording all details of management and production. This may take 2-3 years. On completion of phase III evaluation, final recommendations are made.

D. RANGE CONDITION

Range condition varies considerably depending upon its site flora, climate and use. Stoddart *et al.* (1965) grouped range conditions into five following classes.

Class 1. Climax vegetation, not overgrazed to cause any successional change.

Class 2. Predominantly climax vegetation but invaded by perennial forbs and better annuals.

Class 3. Climax vegetation present but not dominant. Plants with high palatability destroyed.

Class 4. Climax vegetation absent, some valuable plants present, largely non palatable or low valuable vegetation.

Class 5. Climax vegetation absent, land not suitable for grazing because of lack of vegetation or its unpalatability.

Quantitative climax approach for the description of range land has been given by Dyksterhuis (1949) based on percentage of climax vegetation in stand.

Per cent of climax composition	**Category**
0 to 25	Poor
25 to 50	Fair
50 to 75	Good
75 to 100	Excellent

Humphrey (1945) devised site potential approach for classifying range condition based on concurrent production compared to potential based on contribution of key species of known value, plant vigour, seed production, litter and occurrence of soil erosion.

75 to 100 % of potential	Excellent
50 to 75 % of potential	Good
25 to 50 % of potential	Fair
< 25 % potential	Poor

USDA Forest Service Method of range condition classification is based on rating of soil condition, plant density, ground cover, vegetation composition and plant vigour as follows :

1. Soil condition

	Condition	Rating points
Class 1	No soil loss, no erosion	20
Class 2	Soil movement slight, rill erosion noticed, litter not well	17
Class 3	Soil loss noticiable	7
Class 4	Advance stage of erosion, active gullies, exposed plant roots	0

2. Density or ground cover

>0.5	10
0.4	8
0.2	4
0.1	2
<0.1	1

3. Vegetation composition

Class 1	Abundant perennial forage species	10
Class 2	Perennial forages of better quality, moderately abundant	7
Class 3	Perennial forages of better quality, scarce	5
Class 4	Perennial forages of better quality, present as relics	1

4. Plant vigour

This is judged by its colour, size and rate of growth.

Class 1	Palatable plants vigorous, robust, numerous leaves, seed stalks tall and numerous	10
Class 2	Lacking vigour, forage grasses are shorter, fewer seed stalk	7
Class 3	Palatable plants lacking in vigour, grasses weak, forage plants not reproducing	5
Class 4	Palatable plants sickly and weak, non production of palatable plants, sod thinning	1

Rating points of each criteria is totalled up to classify range condition as given below (Figure 26).

Total Points	Category
> 40 points	Good
30 to 40	Fair
15 to 29	Poor
< 15	Very poor

Bhimaya and Ahuja (1969) have used sample air dried forage production as criteria for classification of range condition in Rajasthan.

Air dried forage yield t/ha	Carrying capacity	Condition
1.5	25 - 30	Excellent
1.0	20	Good
0.75	17	Fair
0.50	13	Poor
< 0.2	10	Very Poor

E. DRY MATTER CONTENT

There is a wide variation in moisture content of herbage, thus the yield is expressed on dry matter basis. The dry matter is defined as the constant weight a sample attains when heated at 100°C.

To determine dry matter content of fresh fodder, a sample is taken and dried in oven either for 6 hours at 100°C or till constant weight at 70°C in oven fitted with exhaust fan. Dry weight is recorded to nearest gram. For sampling, if the cut herbage is not more than 15 cm long, one sub sample of 200 g is taken to represent each plot. In case of tall and coarse herbage, chaffed sample of 500 g should be taken from each plot. Drying of fresh sample should preferably be done in trays or perforated coarse paper bags or heat resistant polythene bags. Oven used for drying herbage sample should be of large size, fitted with exhaust fan to regulate hot air and exit of water vapour. Fresh herbage sample come to constant weight in about 6 to 8 hours of drying at 100°C.

Herbage contain active respiratory enzymic systems which continue to function during the early stages of drying and also if sample is not put for drying. This leads to some loss of dry matter till enzymatic activities are put to halt either by denaturation of enzyme or by desiccation. Another limitation in determining true dry matter content of biological material is loss of volatile organic compound which occurs on drying of fresh herbage material. This leads to an under estimation of dry matter content. Drying at low temperature is slow and it may reduce losses of volatile compounds but losses due to prolonged enzymic activity may be higher. Rapid drying at 100°C will reduce loss due to enzymic activity but losses due to volatilization will be higher. Dry matter loss may be in the range of 1 to 2 % under average drying condition. Herbage samples when dried rapidly pickup moisture, thus, must be weighed rapidly once they have cooled.

Faeces dry at a slower rate than herbage because of the formation of an outer crust. Therefore faeces should be broken up and spread out as much as possible before drying. Fresh faeces containing about 12% dry matter may take up to 24 hr at 100°C to reach constant weight. There could be loss of less than 2% of dry matter through volatilization and bacterial action at early stage of drying. For accurate determination of nitrogen, it should be done on fresh faeces.

Silage contain as high as 10% of volatile fatty acid which may be lost on drying at 100°C. Thus, drying of silage should be preferred at lower temperature. Drying of 100 g sample of silage in a flask immersed in boiling water and condensing the volatile fraction for determination of loss of fatty acid on drying can also be done (Anon, 1961).

F. GRASSLAND PRODUCTION

Measurement of pasture production involves two biological systems — the plants and the animals. The quantity of herbage produced by a pasture can be measured by weight, visual estimate and animal output (Ivins, 1959; Mott, 1962; Milner and Hughers, 1968).

The units used for sampling herbage in pasture studies consist of areas, lines or points. The sampling area is frequently determined by a quadrat of square, oblong or circular shape. These units vary in size of 12 in.^{2}, 1.0 yd^2 or 1.0 m^2 and are constructed of metal or wood. Some are flexible for folding. Rings or loops of varying dimensions are used for sampling. The line may also be called a transect. It can be a wire stretched for a given distance over the sward or a visual sighting from one position in the pasture to another position. Herbage may be sampled by point contact also and each provide a measurement. A number of system are used to make point recording e.g. knots on a string, probes positioned along a line and supported by a frame, a spot selected at the toe of a shoe.

Animal production depends upon quantity and continuous supply of forage amount of herbage consumed by the animal, nutritive value and digestibility of the herbage and animals performance. There are a number of methods available for measuring various attributes of grassland and no single one can be designated as being the most appropriate under all condition.

1. Yield measurements

Pasture productivity can be quantitatively measured by weight of herbaceous material (Crowder and Chheda, 1982). It can be used to calculate availability of forage for livestock. It can be determined by:

(a) Cutting and weighing samples from known areas and at appropriate interval of time

(b) Weight estimate based on visual observation

(c) Indirect determination using relationships between selected plant factors and production

2. Weight from clipped plots

Procedure involves cutting of the herbage by hand or mechanically, collecting clippings, weighing fresh material from large plots, drying a sample of known fresh weight and weighing the dried sample. Subsequent cuttings at appropriate intervals are taken and production summed up for the whole season or year production. The dry weight of samples from a given treatment are used to calculate the production of dry matter per unit area (Shaw and Bryn, 1976).

3. The cage method for determining herbage yield

Herbage yield of a permanently grazed pasture is usually accomplished by harvesting areas protected by cages (Figure 27). Cages are randomly distributed throughout the pasture and the herbage within the cage-clipped from time to time during the year. A comparable nearby area is harvested outside the cage at the same time. This allows the determination of herbage consumed by the so called 'cage difference' method. It is assumed that consumption equals the herbage yield from the caged area minus the yield of the outside area. The cage is moved to another area after clipping to avoid the modified environmental effects within the cage.

The total herbage produced per hectare could be calculated by knowing the amount on offer at the beginning of the grazing period that consumed by grazing animals and that uneaten at the end of the grazing period.

The size of the cages has varied with different trials and includes dimensions of 0.91 x 0.91 m, 1.2 x 1.2 m and 1.21 x 2.73 m with a height of 0.45 or 0.60 m. They are usually constructed of heavy wire or small metal rods and are pegged down with metal or wooden stakes. Cage of 1.21 x 2.73 m is used at GRI, Hurley with metal tubing framework covered with mesh chain-link netting. It is constructed as an eclipse about 0.65 m at the centre. Ten to twelve cage must be used.

4. Mower strip method

This consisted of cutting four swaths 0.91 x 4.5 m with sickle or mower just before allowing cattle into the rotationally grazed products. Fresh material was weighed and sampled for dry weight determination. Each pasture was grazed for 10-14 days and rested for 3-4 weeks. The herbage residue after grazing was not considered in the mower strip technique. This method can not be used on continuosly grazed pastures.

5. Indirect methods of estimating herbage yields

Herbage yield is influenced by plant traits or related characteristics such as sward height, density of ground cover or a combination of these. These parameters can be used for estimation of herbage yield. Alexander *et al.* (1962) described a method of incorporating plant height and ground cover for single direct measurement. A light weight plywood piece approximately 0.4 m^2 is dropped on to the pasture sward and distance of each corner is measured from ground level. Height in this case represent percent cover of herbage as well as sward height. The final resting elevation of the board expressed bulk density of sward. Following formula is used to estimate dry matter yield.

$$85\ H - 190 = \text{dry matter (kg/ha)}$$

Where H is the sum of the four corners above ground level. Assuming a total of 50 cm, then 85 x 50 - 190 = 4060 kg/ha of dry matter. This method permits rapid estimation of forage production. Based on this principle, Rising Plate Meter are used in U.K. to measure plant height and density which indicates productivity of grassland. Electron pasture meters are also available for measuring total dry matter yield of grassland.

6. Relative grazing value

To indicate relative grazing value of a grassland a method has been devised by Bibby *et al.* (1982). In this method, all the plant species present in the grassland are recorded on 0-8 scale on the basis of their dry matter productivity but also taking into account their sward forming ability, season variation and palatability. The rating of each species is multiplied by its percentage area cover and the products are totalled to get the relative grazing value which is grouped into following three broad categories :

High	=	> 500
Moderate	=	200 - 500
Low	=	< 200

Thus, both presence of desirable species as well as its proportion would determine the relative grazing value of grassland.

A grassland with only a few species having high productivity and high percentage area cover would fall in high grazing value category, whereas, if their proportion is very low, it would fall in low category.

7. Live weight gain

Live weight gain per animal per unit area is determined to assess production and the feeding value of grassland. It is also used for the comparison of grazing systems. Its determination gives true estimate of herbage production and its utilization and takes into account nutritive value and digestibility of herbage. For estimation of live weight gain, generally young animals are used because an adequate quantity of live weight gain is achieved in a relatively short time. Where the live weight gained over the experimental period is the main result, the key weighings are obviously those measuring the initial and final weights of the animal (Anon., 1961). The weight of 'fill' (the contents of stomachs and intestines) in test animals varies with the quantity and quality of feed available, time spent for grazing, and drinking of water just before weighing and it may bias the result. To keep this error to a minimum, the day and time for weighing are chosen keeping in view the behaviour pattern of the stock and the feed available. For determining live weight gain following a particular grazing treatment, initial weight of stock is recorded. It is also used to allocate stock to treatments by weight. Whether the stock are at pasture or in yards, they are kept in one group, or given feed from the same source in separate groups in order to provide uniform nutrition and 'fill' conditions. Two or three weighings are recorded and the initial weight is an average of these.

Usually weight is recorded 3-4 hours after sunrise that is between 9 and 11 a.m., and feeding or grazing is withheld until after weighing. Intermediate weighing during course of study is also carried out to see the trend. It is important for the establishment of growth curves and to express the gain per day.

Monthly intermediate weighings are often adequate for summer grazing experiments, but fortnightly weighing is practised for trial over shorter periods. In certain critical short period trial, stock are weighed daily for 5-6 days. Final weight is determined on two or three successive days. If the treatments are likely to produce marked differences in 'fill', the stock are grazed on a common paddock for two or three days before weighing.

G. SEED TREATMENT

1. Scarification

Seeds particularly of tropical legumes have hard seed coats making them impermeable to water imbition which results in low germination. Scarification is required to improve their germination. It can be done by any of the following means.

i. Acid treatment : Seed is put in concentrated sulphuric acid for 3-5 minutes, followed by thorough washing and drying. Acid resistant plastic bags are useful in enclosing seeds for submerging in the acid and subsequent washing. Seeds of a number of grasses also respond to scarification.

ii. Hot water : Soaking of seed in hot water at varying degrees of temperature and for different duration depending upon species of legumes and grasses, also improves germination. Usually water is heated to about 90°C and removed from the source of heat and then seed is put into it and allowed to remain in it until the water temperature drops to 40°C, then removed, cooled and dried.

iii. Mechanical : Handrubbing of seed with sand paper or carborundum paper causes some abrasion to seed and improves germination of different legumes. In this process, some seeds remain unscarified and results in low germination. Some seed scarifying equipments which consist of gear-operated baffles operating within a cylinder or drum lined with an abrasive are also used. Precautions are required to be taken in using these equipments so that there is not much damage to seeds by cracking or removal of the seed coat due to over scarification (Crowder and Chheda, 1982).

2. Inoculation

There are a diverse group of nitrogen fixing bacteria that establish symbiotic relationships with legumes. These bacteria fix atmospheric nitrogen (N_2) to a metabolizable active form ammonia which in turn can then be incorporated by means of other enzymes into cellular protoplasm. The potential amount of nitrogen fixed by these bacteria depends heavily on the ecosystem in which the organisms operate. Amounts of nitrogen fixed range from trace amounts for some free living soil bacteria to 850 kg/ha/yr (Mengal and Kirky, 1979). Usually the nitrogen fixation is in the range of 100-150 kg/ha. Among nitrogen fixing bacteria *viz. Rhizobium, Azospirillum* and *Azotobacter,* the most widely used inoculant is *Rhizobium*. Nitrogen fixing systems offer an economically attractive and ecologically sound means of reducing external inputs and improving internal resources. Symbiotic systems such as that of legumes and *Rhizobium* can be a major source of N in most cropping systems. Nitrogen fixation by association and free living microorganisms can also be important (Peoples and Craswell, 1992).

In addition to microbial inoculants of N_2 fixing bacteria, vesicular arbuscular mycorrhizae (VAM), phosphate solubilizers and plant-growth promoting rhizobacteria (PGPR) are also important in improving nutrient supplies to plants. In mycorrhizae, there is symbiotic association between plant roots and fungal mycelia of *Endogone, Glomus, Sclerocystis Entrophosphora, Acaulospora, Gigaspora* and *Scutellispora*. Rhizobacteria that exert a beneficial effect on plant growth are referred as plant growth promoting rhizobacteria or PGPR. These improve plant growth by colonizing the root-system and preempting the establishment of or suppressing deleterious rhizosphere microorganisms (DRMO). Rhizobacteria belong to several genera, e.g. *Azotobacter, Bacillus, Pseudomonas, Rhizobium, Streptomyces, Xanthomonas, Actinoplanes* and *Agrobacterium*. Inoculation of pasture legume seed with a suitable *Rhizobium* culture is necessary for their better establishment.

It is estimated that about 139 million tonnes of N_2 is biologically fixed in various terrestial system out of which 32.3% is in permanent grasslands.

The major species of Rhizobium are listed in Table-6.

Table 6. The major species of *Rhizobium*, cross incoulation group and host genera

Rhizobium species	Cross-inoculation group	Host genera
R. meliloti	Alfalfa group	*Medicago, Melilotus, Trigonella*
R. trifolii	Clover group	*Trifolium*
R. leguminosarum	Pea group	*Pisum, Vicia, Lathyrus, Lens*
R. phaseoli	Bean group	*Phaseolus*
R. lupini	Lupine group	*Lupinus, Ornithopus*
R. japonicum	Soybean group	*Glycine*
Rhizobium sp.	Cowpea group	*Vigna, Cajanus, Cicer Lespedeza, Crotolaria, Pueraria*

Legumes differ in their needs for particular bacterial strains. Some will nodulate from a very wide range of native and selected strains. Three categories of species are recognized depending upon their promiscuity or degree of selectiveness.

A. Unselective : Calapo, Cowpeas, Dolichos, Siratro, Stylo, Puero, Velvet beans, Teramnus

B. Moderately Selective : Centro, Desmodium, Glycine

C. Highly selective : Clovers, Lotononis, Leucaena, Lucerne, Medics, Peas, Vetches

Various forms of inoculant are available such as agar culture, freeze-dried or peat slurry, but the peat inoculant is the most widely used because its survival on seed is much better than other forms.

Usually 250 g of inoculum and one litre of water is normally required to treat every 25 kg of clover seed. Dusting of peat culture on the seed and mixing it before seeding is a simple way of inoculation. However, for better and more efficient inoculation, a 10% selection of jaggery (*Gur*) mixed with *Rhizobium* culture should be used for inoculating the seed. Inoculated seed may be dried in the shade before seeding and should not be exposed to sun to avoid mortality of the *Rhizobium* bacteria present in the culture. The treated seed should be spread out in a thin layer to dry in a cool place indoors out of direct sunlight which kill the bacteria. Once inoculated the viability of rhizobia can be quite short, therefore, sowing should be done within 24 hours and not later than 48 hours. Close contact between the inoculated seed and fertilizers should be avoided. Lime and basic super phosphate can be mixed with treated seed. Usually 1:15 ratio of sticker to seed is used for inoculating seed.

Tropical legumes once considered inefficient as fixer of atmospheric nitrogen are now known to fix amounts of nitrogen equivalent to most temperate legumes, but these amounts of nitrogen (40-260 kg N/ha/yr) are usually well below the 300-1000 kg N/ha/yr needed to sustain the high growth of the grasses. Consequently some tropical grasses are frequently deficient in nitrogen and their crude protein contents are low (Wilson, 1984; Tilak and Singh, 1996).

3. Pelleting

Seeds of most of forage grasses and legumes are very small in size and light in weight. When sown there is risk of their being washed or blown away. In their seeding in rangeland or areas having grass turf, these do not even get chance to come in contact with soil preventing there germination and establishment.

To facilitate better establishment of grasses and legumes particularly in rangeland, native grassland and in rough areas vulnerable to erosion, seeding of pellatized seed is done. Seed pelleting is the coating of grasses and legumes with a layer of fine powder held on to the seed by a sticker. This aids in sowing it into relatively unfavourable soil condition, establishment and effective nodulation of legumes particulary in temperate legumes. Lime is commonly used as a coating agent for temperate legumes. It is not a appropriate coating material for pelleting tropical legume seed except for *Leucaena*. Since phosphorus is usually the limiting factor in establishment of tropical legume, rock phosphate which has been ground to pass a 300 mesh sieve should be used for pelleting tropical legume seeds (Crowder and Chheda, 1982). Methyl cellulose (4%) and gum arabic (40%) are commonly used as adhesive in seed pelleting.

The pelleting process consists of following :

a) Preparation of adhesive by mixing with hot water and then cooling

b) Addition of inoculant to the adhesive

c) Tumbling of the seed and adhesive- inoculant solution in a barrel, concrete mixture or bucket

d) Addition of pelleting material with continued tumbling until seeds are evenly coated and separated

e) Drying of the pelleted seeds

Various seeds require different quantities of adhesive and pelleting material because of size variation. For pelleting seed, seeds can be added in mixture of sand, clay, compost and fertilizer in ratio of 3:1:1:1 and using sufficient quantity of water. This mixture is put in shaking drums like seed treating and tumbled for few minutes. Pellets or small balls of convenient size are prepared in such a way that each pellet contains 2-3 seed. The pellets are dried and stored for 4 to 6 months before sowing. Pelleted seeding gives the seedling a

Fig. 24. Rye grass and white clover strains being evaluated under spaced planting

Fig. 25. Rye grass and white clover strains being evaluated under sward condition

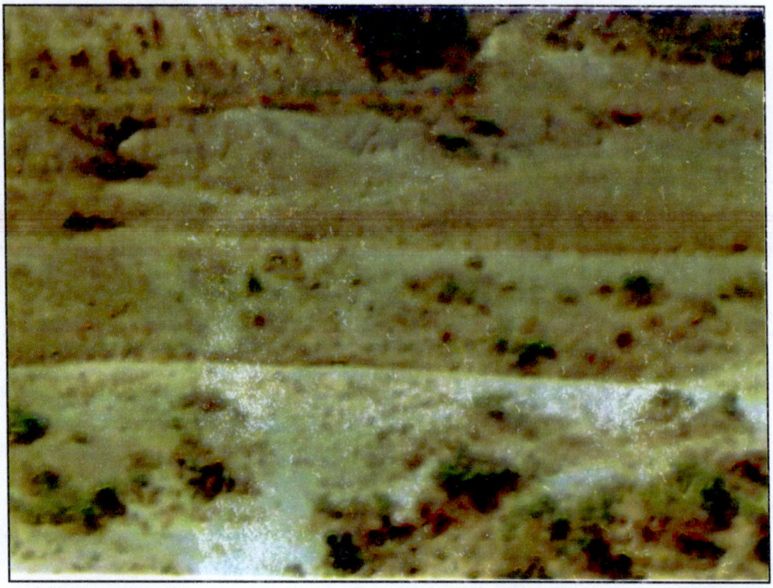

Fig. 26. A very poor, overgrazed and degraded grassland in Shivalik

Fig. 27. Iron mesh cages in grassland to exclude grazing for determination of herbage yield and botanical composition

vigorous start and also chance to establish better particularly in unprepared soil and rough areas prone to erosion.

The sowing of pellets are carried out before premonsoon showers in plains or at the onset of spring season in high hills. Sowing rate varies with the forage species and the condition of land where seeding is required to be done. The pelletized seed has disadvantages of being expensive and sometime reducing viability (Lal, 1990).

Dolomite, Gofsa phosphate, and Nauru phosphate are also used as pelleting material. Gum arabic adhesive and fine lime for pelleting seed should be preferred as these materials maintain high populations of rhizobia on the pelleting seed than the more commonly recommended rock phosphate and dolomite pellets with methyl cellulose adhesive. The age of commercially coated seed should be checked, and if it has been manufactured for more than one month it should be rejected. There is increase of 60-80% in weight of pelleted seed and allowance should be made for this in seed rate. Pelleting with lime alone increase nodulation to 60.4 as compared 0.8% in simple inoculated seed of lucerne. The ratio of pelleting material and seed is usually 2:3.

H. CARRYING CAPACITY

Number of animals that can graze in an unit area without over or under grazing in an average season is referred to as carrying capacity of grassland. It can be assessed by actual grazing of animal or by determining dry matter production from enclosed area and relating it with actual dry matter requirement of animal (Figure 27).

Natural grasslands have vegetative growth only for 3-5 months in a year due to moisture or temperature limitation. The carrying capacity of grassland is expressed in terms of livestock unit per hectare.

Dry matter herbage production can be estimated with the help of one meter square quadrat. All the palatable species inside the quadrat are clipped to 2.5 cm stubble height. For browse the young twigs are clipped upto a browsable height. The clipped material is dried in oven and dry weight recorded.

The clipping of forage grasses and browse is done as and when preflowering or sufficient growth is observed. The dry weight of all clipping is summed up and the average production of all the quadrats of different range sites are calculated. The forage production of each range site is expressed in kg/ha. Grazing capacities for each range site are calculated and expressed in terms of animal unit days per hectare. Grazing capacity is calculated at the rate of 2 kg dry matter/sheep unit/day at 60 per cent utilization level.

Another way to determine carrying capacity of a rangeland is to have an assessment of the total annual forage production of the rangeland and divide it by the average annual requirement of the livestock. The grazing capacity per hectare on annual basis can be expressed as follows :

$$\text{Grazing capacity} = \frac{\text{Total annual forage production}}{\text{Forage requirement animal}^{-1} \times 365 \times A}$$

where A is the area of grazing land.

A method adopted in U.K. for computation of carrying capacity is as follows (Anon., 1980). If 50 sheep are grazed for four months on a rough grassland area of 10 hectare, the carrying capacity can be determined as follows :

Live Stock Units

5 sheep	=	1 L S U
50 sheep	=	10 L S U

Rough grazing on 10 ha for 4 months

$$\frac{10 \times 4}{12} = 3.3 \text{ L S Us}$$

$$\text{Carrying capacity of rough grazing} = \frac{3.3}{10} = 0.33 \text{ LSU/ha}$$

1. Ocular reconnaissance method

In this method, an assessment is made in respect of each plant species of the rangeland which are likely to be grazed and their percentage composition recorded. Forage acre factor and forage acre values are computed.

Forage acre factor =

Summation (percentage species composition x use factor of each species) x Forage density of the type

Forage acre = Forage acre factor x Surface acres

The forage acres of the rangeland divided by the forage acre requirement per animal unit will give the carrying capacity of the range land in terms of animal unit that the rangeland can support.

The method has limitation as there is no definite relationship between forage index and density. The forage will vary with the growth habit of the plant species.

2. Forage use factor method

Carrying capacity of a grassland can be estimated as per methodology advocated by Pandey (1981) and Joshi *et al.*(1990).

The outline of methodology is as follows.

One protected plot of 50 m x 50 m is selected for the monthly forage production. One plot of 50 m x 50 m is demarcated as the continuous grazing plot. Biomass studies from both the plots are done at monthly interval. Carrying capacity studies can be done for any type of animal like cow, buffalo, horse, sheep and goat. Bite count, bite size, time of each visit and number of visits per day are recorded for each cattle group. Besides this, total grazing hours per day, percentage bite estimates are also recorded.

a) Hours spent grazing/day = (Grazing hours/day) - (Time spent on walking + Time spent on resting + Time spent on other activities)

b) Bite Count = Number of browsing sounds produced in one minute

c) Bite size = $\dfrac{\text{Wt. of intact Patch} - \text{Wt. of grazed patch}}{\text{Number of bites on the grazed patch}}$

d) Forage dry matter consumption/animal/day
= Bite size x Total bite numbers/day

e) No. of visits/day x Time spent in grazing/visit/2500m^2
= Total grazing period/2500m^2/day

f) Consumption/2500 m^2/day =

$$\frac{\text{Consumption/animal/day}}{\text{Time spent for grazing/day}} \times \text{Total grazing period/2500 m}^2\text{/day}$$

g) Consumption/m^2/day = $\dfrac{f}{2500\ m^2}$

h) Consumption/m^2/month= g x 30 (days of month)

Standardization of different animal groups against cow :

$$\frac{\text{Food consumption of cow/day}}{\text{Food consumption of particular animal group/day}} = \text{No. of cows}$$

Spedding (1971) described the following formula to assess degree of herbage exploitation :

$$\text{Herbage exploitation} = \frac{\text{Total consumption/m}^2}{\text{Total above ground production/m}^2} \times 100$$

3. Proper forage use factor

The proper forage use factor is the percentage use that is made of a forage species under proper management. It is expressed as the percentage that is consumed of the current years forage production of a particular species which is within reach of stock. The use occurring on overgrazed range is not currently the proper factor of species. According to Stoddart *et al.* (1965), the proper use factor is the cumulative utilization of a species when full (correct) use has been made of a range. It is a product of preference and quantity available.

The proper use factor of the forage is determined by an experiment carried out in 6 fenced plots of 20m x 20m area. Each plot is further fenced into 6 smaller plots of 1m x 1m size. Each month a sample plot of 20m x 20m size is opened and grazed to its maximum utilization without affecting the animal diet, that is animal is provided with extra fodder to complete his diet. A minimum forage supply of 2.4 kg/100 kg/day body weight was taken into consideration with digestive efficiency of 48%. On each sampling date, the vegetation inside the 6 plots of 1m x 1m size (Intact patch) is clipped. Six more plots of 1m x 1m size adjacent to the fenced ones are harvested. These represented the grazed patch. The difference in the dry weight of intact patch-grazed patch when averaged, gives the proper use factor of the forage of the respective month. Similar experiments are done in the coming months with a new sampling plot of 20m x 20m size. The proper forage use in the study site calculated in a grassland in Garhwal region was 48%.

$$\text{Carrying capacity (animal/ha)} = \frac{\text{Total forage production/ha} \times \text{Proper forage use factor}}{\text{Animal requirement}}$$

Gupta (1978) estimated some 2,800,000 cattle and 1,100,000 small stock in U.P. Himalayan district giving flock densities of 70 and 27 km^2. Capacity of 0.35 to 0.90 cow/ha assuming 50% herbage utilization has been reported by Sundriyal (1989) from alpine pastures in Garhwal Himalayas. Phillimore (1984) reported stock densities of 120 km^2 in Kangra district of H.P. The carrying capacity of an alpine grassland located at an altitude of about 4000 m in Garhwal Himalayas has been found to vary from 0.6 to 1.5 cows/ha with potential carrying from 1.3 to 3.4 cows/ha on the basis of 48% herbage utilization. The grazing pressure has been worked out to be 10 times of the existing carrying capacity (Joshi *et al.*, 1990). For a temperate pasture in Himachal Pradesh carrying capacity of only 2 sheep/ha per annum has been reported (Dogra *et al.,* 1994). For the grazing season, it works out to be 6 sheep/ha. The present level of carrying capacity of temperate and alpine pasture is low due to short growing season, lack of proper management and high grazing pressure. The studies conducted on different aspects of pasture indicate that carrying capacity of temperate pastures could be increased to 15 to 20 sheep/ha in an average growing season if pasture are improved and properly managed.

I. LIVESTOCK UNIT

The livestock unit is a concept developed to equate livestock numbers with the carrying capacity of land. It has come into fairly general use and is particulary helpful where precise recording of land utilization is not carried out. It is based on the assessment of dietary energy needs of various types of livestock (Anon., 1980). Thus different classes of animals at varying ages and levels of production can be related to a standard live stock unit (L S U).

It is useful in comparative analysis for assessing stocking intensity within one farming system as compared with others. L S U is a standardized animal unit to which different ages, types or species of livestock can be related for purposes of matching forage availability to animal needs or comparing different livestock enterprises or different livestock type (Lal, 1990).

An animal unit is widely accepted as a mature cow with calf of 1000 lb (450 kg) liveweight, the other animals are converted to animal unit. About 20 lb of dry fodder is needed by 1000 lb (450 kg) liveweight cattle.

Ahlgren (1956) proposed following units for different types of livestock :

Dairy cow (1000 lb)	1.0 unit
Beef cow	1.0 unit
Dairy heifer	0.4 unit
Beef steers	0.4 unit
Beef heifers	0.4 unit
Horse or mule	0.75 unit
Ewe	0.2 unit

Each animal unit consumes the equivalent of 3 tons of dry forage per year.

Shultis (1962) described animal units as follows :

2 year old beef cow, bull, steer or heifer	1.0 unit

Yearling steer or heifer weighing 275-320 kg	0.75 unit
Weaner calf (8 or 9 months of age) 160-200 kg	0.5 unit
Calf 4-8 months	0.3 unit
A ewe, ram or wether over 1 yr.	0.2 unit
A lamb up to 1 yr.	0.15 unit

Moorhouse (1970) expressed relationship between different classes of livestock and animal unit as follows :

400 kg Steer	=	1 animal unit
Breeding cow	=	2.0 au
Calf 1-8 month	=	0.35 au

Horses, sheep and goats are commonly converted to animal unit at the rates of 1.25, 0.2 and 0.17, respectively (Heady, 1975).

The H.P. Grazing Advisory Committee recommended following animal units for different categories of animals (Anon., 1970) :

Buffalo	-	6 unit
Cattle	-	4 unit
Sheep	-	1 unit
Goat	-	1.5 unit
Horse	-	4 unit
Mule	-	5 unit
Donkey	-	3 unit
Yak/Churu	-	4 unit

Animal units of cattle, buffalo, sheep and goats have been equated on a relativity of 0.8, 1.0, 0.1 and 0.1, respectively (F.A.O. 1979). The Live Stock Unit coefficients in use for cattle and sheep in United Kingdom are as follows (Anon., 1980).

Dairy cows	1.00
Dairy bulls	0.65

Beef cow	0.75
Beef bulls	0.65
Other cattle	
0-12 months	0.30
12-24 months	0.54
Over 24 months	0.80
Sheep Light weight (40 kg)	0.07
Medium weight (60 kg)	0.09
Heavy weight (80 kg)	0.11
Rams	0.08
Lamb	0.04

Sardar (1989) proposed animal unit as follows :

Cattle	-	1.0 unit
Buffaloes	-	1.5 unit
Sheep	-	0.2 unit
Goats	-	0.3 unit
Camels	-	1.7 unit
Donkeys	-	1.0 unit
Horses	-	1.3 unit

J. SOIL MOISTURE

The moisture has the greatest influence on the growth and yield of forage grasses and legumes. The moisture content of soil is expressed in terms of percentage based on the weight of soil. It is determined by weighing a sample of the moist soil, drying it at a temperatue of 105°-110°C and reweighing it when cool. The attraction between soil and water is so great that all the moisture is not removed by drying at a temperature of 110°C although the quantity of moisture remaining is very small and a soil so dried is considered to be moisture free for general purposes. The amount of moisture in soil which occurs as a thin film (4-5 milli micron) and is held tenaciously with a tension of 31 atmosphere or more is known as hygroscopic moisture. Water held in this condition does not move in the soil, is not used by plants and in fact may exist in other than the liquid state. Water held by a soil in excess of the hygroscopic coefficient is considered to exist in the form of thin film around soil particles and granules and is called capillary moisture which is susceptible to movement in the soil and to a certain extent may be absorbed by plant roots. Capillary water is held at a surface tension ranging from 1/3 to 31 atmosphere. It is the only important source of water for most cultivated plants.

Tensiometers are used to measure the tension or capillary potential of the soil moisture. They consist essentially of porous porcelain cups, which are filled with water and buried in the soil and are connected by tubing to mercury, manometers or vacuum gauges for indicating the pressure or tension, on the system when the water in the cup is in equilibrium with the water in the soil. If tensiometer is installed below the surface of the water table, it will show positive pressure. In saturated soil under no hydrostatic pressure, it will show zero pressure, and in unsaturated soil it will show negative pressure or tension.

Water which occupies the larger pores of the soil and drains away under the influence of gravity is called gravitational water. It is free water held at a tension below 1/3 atmosphere and is often injurious to plants if drainage is too slow.

The percentage of water retained by a soil when it has been subjected to a centrifugal force of 1,000 times gravity in a soil centrifuge is referred as moisture equivalent. The normal field capacity, normal moisture capacity and field carrying capacity refers to the percentage of moisture in a soil after drainage has ceased and after capillary adjustments have taken place. In most cases, it represents the water held at tension of 1/3 atmosphere.

It is generally recognized that capillary water held at tension greater than 15 atmosphere, is not available to plants. At this point of soil moisture, the plant wilts permanently.

The moisture content of the soil at this point when the leaves of plants growing in that soil first become permanently wilted has been variously designated as the wilting point, wilting coefficient, wilting percentage and permanent wilting percentage.

The available water content of field soil is defined as the amount of water that a crop can remove from the soil before its yield is seriously affected by drought. It is also defined as the amount of water in a soil held between its condition at field capacity and at its permanent wilting point, that is the moisture held at tension between 1/3 atmosphere and 15 atmosphere.

As the quantity of moisture in the soil becomes less, the force by which it is retained increases. Finally, a point is reached at which the adhesion between soil and water equals the absorptive power of the plant roots, and hence no more moisture can be utilized by the plant, and this point represents lower limit of available moisture. The upper limit of available water is assumed to correspond roughly with the percentage of water retained (Millar *et al.,* 1958).

Water holding capacity of soil of grassland plays an important role in determining sward yield. It is expressed as available water capacity (AWC) which depends upon soil texture, stoniness, organic matter content, structure and depth to rock or other impediments to rooting. Based on AWC, soils can be divided into following three categories (Anon., 1982).

1. Low AWC (less than 100 mm) - Soils consisting of sands and gravels or shallow stony soils with rock at less than 40 cm
2. Medium AWC (100-150 mm)- Clay soils with flints over chalk, loamy soils with rock or gravel at 50-80 cm, clay soils and shallow loams over clay
3. High AWC (more than 150 mm) - The soils in this category consist of deep organic and peaty soils and well drained loamy soils over 1 meter deep

K. SOIL TESTING

Soil testing is an aid in determining the fertility level of soil. It shows organic carbon, nitrate, nitrogen, available phosphorus and potash, soil pH (acidity or basicity), soluble salts (salinity), lime content and texture of soil. Proper sampling of soil for soil testing is very important and following procedure need to be adopted.

A. Time of sampling : Sampling should be done before the seeding of crop or before the time of fertilizer application as in case of grasslands.

B. Sampling depths : For all crops, it is necessary to sample and analyze the soil at two separate depths *viz* 0-15 cm and 15-60 cm. However, for convenience, commonly samples are taken from 0 - 15 cm. depth.

C. Sampling procedure : The soil samples submitted from each field must be representative of that field. Four to five samples should be taken from an area of an acre. These samples should be mixed and then two representative samples weighing about 0.5 kg should be taken for analysis. Sample should be put in cloth bags with label giving the address of land owner, field location and number, sampling date etc. Farmlanes, field borders, dead furrows, old manure piles or old straw stick bottoms, eroded knolls, poorly drained areas should not be sampled. If soil samples are too wet, these should be dried using fan or in shade. Soils are dried at 30°C in a fan ventilated oven. It is spread in a thin layer in shallow trays to minimize any incubation effect due to slow drying. After four days in the drying oven, the soil is stored in waxed card board cartons for analysis. For sampling of soil, first grass or crop residue on upper surface of soil should be removed and then using spade, hand hoe (Khurpa) or augar soil from top to 15 cm depth be taken through a V shape digging.

Soil test report

Soil test report usually contains information on organic carbon percentage, available nitrogen, phosphorus and potash content (kg/ha) expressed in low, medium and high group which indicates values as given in Table 7.

Table 7. Major soil nutrients and their grouping according to values.

Constituent	Low	Medium	High
Organic carbon %	< 0.5	0.5 - 0.75	> 0.75
N kg / ha	< 280	280 - 560	> 560
P kg / ha	< 10	10 - 25	> 25
K_2O kg / ha	< 110	110-280	> 280
pH	< 7 (Acid)	7 (Neutral)	> 7 (Alkaline)

Total soluble salts – Degree of salinity / alkalinity

Texture - varying from clay to loam sand

Recommendations of nutrient requirement of crops in various package of practices are made for medium soil and are increased and decreased by 25 % for soils with low and high level of nutrients respectively. For pH following categorization is done.

pH	**Reaction**
> 8.5	Alkaline
7.5-8.5	Saline
6.5-7.5	Near Neutral
< 6.5	Acidic

Total soluble salts as determined by electrical conductivity indicates the degree of salinity, alkalinity etc. of the soil. It is expressed in millimhos/cm and categorized in following four groups.

(a) < 1 = normal

(b) 1 to 2 = critical for germination

(c) 2 to 4 = critical for growth of the sensitive crops

(d) > 4 = Injurious to most crops

4

GRASSLAND MANAGEMENT

The basic principle of grassland management is to improve and maintain grasslands for higher production of fodder rich in nutritive value and palatability. Grassland or range management involves proper management of basic resources which may include soil, water, vegetation and animals to have optimum level of nutrient production per unit area (Lush, 1952; James, 1969). It is observed that practically no grassland improvement or management practices are being adopted in our grasslands except for some control on grazing and enclosure in certain areas. Because of this productivity of our grassland is very low (Whyte, 1964). However, it could be increased considerably over the present level, if suitable measures for their improvement and better management are adopted. This includes introduction of legumes, reseeding of grasses, optimum fertilizer use, removal of undesirable bushes, plantation of browse and proper cutting and grazing management and adoption of soil and water conservation practices. Measures to be adopted for their management depend upon the condition of grassland and financial resources at hand (Archer and Bunch, 1953; O'Reilly, 1975).

A. RESEEDING OF GRASSES

High grazing pressure over long period of time is primarily responsible for elimination of desirable species and deterioration of grassland. The continuous heavy grazing may lead to destruction of vegetative cover and inturn severe soil erosion. Prevention of grazing induces progressive succession and improvement in ground cover by grasses and other species. Closure of grassland for a few years will be very effective if remnants of desirable species are still present to serve as seed source for regeneration (Figure 28). In higher rainfall areas, elimination of grazing in grassland may lead to regeneration of bushes, weeds and trees also if grassland is not the vegetation climax.

If the grassland is in very poor condition, regeneration on closure shall be very slow. In such situation, reseeding would result in faster regeneration. Selection of a species for seeding keeping in view its adaptability to a particular area is very important (Singh and Katoch, 1975; Singh *et al.* 1981; Singh and Sharma, 1993). Broad outline of adaptability of different grasses depending upon altitude, rainfall and specific adaptability is given in Table 8 (Hitchocock, 1950; Barnard, 1964; Langer, 1993; Humphreys, 1998; Miller 1984; Rehm and Espig, 1991).

The seeding of grasses can be done, after ploughing and complete seed bed preparation or by over sowing, sod seeding (over drilling) or hay mulch seeding. Seeding after ploughing and complete bed preparation is expensive and is suitable only in plain or area with moderate slope only (Figure 29). Introducing or reintroducing desirable plant species to a rangeland may be advantageous over complete seed bed preparation where erosion hazard is high, preparation of seed bed is impractical if purpose is to modify rather than replacement of the present plant stand. Introduction of setaria (*Setaria anceps*) cv. PSS1 and guinea grass (*Panicum maximum*) cv. PGG 9 and PGG 19 have been tried in a natural grassland of mid hills in H.P. Introduction through seeding or planting of slips in pits has been found to be superior to scrapping or killing existing vegetation with paraquat spray in band and then seeding (Sood and Kumar, 1995; Kumar and Sood, 1998).

The details of seed count, seed standard and seed rate of different grasses are given in Table 9 (Wheeler and Hill, 1957; S'kerman and Riveros, 1990). Over sowing or sod seeding (over drilling) is convenient for seeding in grasslands. Surface broadcasting of seed into an existing grassland is referred as oversowing. It is preferred in grasslands which are too steep or rocky. In sod seeding or over drilling, seed is drilled into existing grassland without previous cultivation. Overdrilling is more reliable than oversowing. Another convenient method of reseeding of grass in grassland is hay mulch seeding in which hay is harvested when seed is nearly ripe. Hay is then spread by hand in grassland, gullies, dams, spillways, waterways,

Table 8. Adaptability of certain grasses under different agroclimatic conditions

Grass	Agroclimatic condition					Special Adaptability
	Tropical subtropical (<800 m)	Subtemperate (800 - 1800 m)	Temperate wet (1800-4000 m)	Temperate dry (2400-5000 m)	Minimum Rainfall (mm)	
1	2	3	4	5	6	7
Black spear grass (*Heteropogon contortus*)	+++	+	–	–	600	Drought
Buffel grass (*Cenchrus ciliaris*)	+++	+	–	–	350	Drought
Blue panic (*Panicum antidotale*)	+++	+	–	–	450	Low fertility
Dinanath grass (*Pennisetum pedicellatum*)	+++	+	–	–	600	Sand
Guinea grass (*Panicum maximum*)	+++	+	–	–	1000	Moderate fertility
Marvel grass (*Dichanthium annulatum*)	+++	+	–	–	600	Grazing
Napier grass (*Pennisetum purpureum*)	+++	+	–	–	750	Cut & Carry
Dhaulu grass (*Chrysopogon fulvus*)	+++	+++	–	–	500	Low fertility
Love grass (*Eragrostis curvula*)	+++	+	–	–	600	Drought

1	2	3	4	5	6	7
Rhode grass (*Chloris gayana*)	+++	+	–	–	650	Drought
Sadabahar (*Andropogon gayanus*)	+++	+	+	–	600	Drought
Sabi grass (*Urochloa mosamibicensis*)	+++	+	+	–	400	Grazing
White grass (*Sehima nervosum*)	+++	+	–	–	400	Drought
Vetiver* (*Vetiveria zizanioides*)	+++	+++	+	–	500	Soil conservation
Bhabbar grass* (*Eulaliopsis binata*)	+++	+	–	–	500	Soil conservation
Signal grass (*Brachiaria decumbens*)	+++	+	–	–	1000	Grazing
Sewan grass (*Lasiurus hirsutus*)	+++	–	–	–	250	Grazing
Giant Star grass (*Cynodon plectostachayus*)	+	+++	–	–	600	Water ways
Para grass (*Brachiaria mutica*)	+++	+	–	–	1250	Water logging
Lunji (*Themeda anathera*)	+	+++	+	–	600	Low fertility
Makarikari grass (*Panicum coloratum*)	+	+++	+	–	500	Drought

1	2	3	4	5	6	7
Pangola grass (*Digitaria decumbens*)	+	+++	–	–	600	Shade
Broad leaf paspalum (*Paspalum wettsteinii*)	+	+++	+	–	1000	Grazing
Dallis grass (*Paspalum dilatatum*)	+	+++	+	–	900	Grazing
Bari Jhan (*Pennisetum orientale*)	+	+++	+	–	750	Soil binding
Setaria (*Setaria anceps*)	+	+++	+	–	750	High rain fall
Molasses grass (*Melinis minutiflora*)	+	+++	–	–	750	High rain fall
Tall fescue (*Festuca arundinacea*)	–	+	+++	+	500	Drought
Meadow fescue (*Festuca pratense*)	–	+	+++	+	500	Grazing
Phalaris (*Phalaris tuberosa*)	–	+	+++	+	500	Drought
Orchard grass (*Dactylis glomerata*)	–	+	+++	+	600	Light shade
Crested wheat grass (*Agropyron cristatum*)	–	+	+	+++	400	Drought cold
Crested dogs tail (*Cyanosurus cristatus*)	–	–	+++	+	500	Drought

1	2	3	4	5	6	7
Intermediate wheat grass (*Agropyron intermedium*)	–	+	+	+++	400	Drought cold
Kentucky blue grass (*Poa pratensis*)	–	–	+++	+	500	Grazing
Prairie grass (*Bromus catharticus*)	–	–	+++	+	500	Drought grazing
Smooth Brome grass (*Bromus inermis*)	–	–	+++	+	500	Drought
Timothy (*Phleum pratense*)	–	–	+++	+	1000	Moist
Rye grass (*Lolium perenne*)	–	–	+++	+	800	High fertility
Italian rye grass (*Lolium multiflorum*)	–	–	+++	+	800	High fertility
Red top (*Agrostis alba*)	–	–	+++	+	600	Wet soil

+++ = Good, ++ = Average, + = Below average, – = Not adapted
* = Fodder value only at early growth stage

dunes and sand blow outs. To facilitate easy establishment of sown species, elimination of competition of existing flora is desirable. It can be done by discing and harrowing in grassland which are plain or have negotiable slope. The competition from existing flora can also be reduced by use of herbicide like gramoxone in bands at suitable interval.

Table 9. Seed count, seed standard and seed rate for common grasses

Tropical Grasses

Grass	Seed count per kg, 000	Minimum purity, %	Minimum germination	Seed rate, kg/ha
Birdwood grass	500	90	20	4
Blue panic	1500	80	50	5
Buffel grass	500	90	20	4
Dallis grass	700	60	60	10
Dinanath grass	500	90	20	4
Dhaulu grass	1300	60	20	5
Guinea grass	5000	60	20	6
Hybrid grass	–	–	–	25000 slips
Kikuyu	40	93	60	5*
Klein grass	1200	80	20	10
Love grass	3850	90	40	2
Lunji grass	650	60	20	20
Marvel grass	1800	60	20	5
Molasses grass	6000	40	30	5
Para grass	1000	40	15	5*
Rhode grass	4000	90	20	5
Sabi grass	800	60	3	6
Setaria	1200	60	20	5
Sewan grass	600	60	20	10
Signal grass	300	50	15	3
White grass	800	60	20	12

* Propagated easily by vegetative means

Temperate Grasses

Grass	Seed count per kg, 000	Minimum purity, %	Minimum germination, %	Seed rate, kg/ha
Bent grass	20,000	95	60	10
Blue grass	5,000	85	80	10
Brome	120	95	85	10
Crested wheat grass	450	95	80	10
Fescue meadow	500	95	80	10
Fescue tall	450	98	50	6
Harding grass	750	98	60	4
Intermediate wheat grass	220	90	80	10
Orchard grass	1000	90	80	10
Prairie grass	150	95	85	10
Red fescue	1000	95	80	3
Reed canary grass	1300	98	60	6
Rye grass Italian	400	96	75	6
Rye grass perennial	500	96	80	4
Timothy	2600	95	80	5

B. INTRODUCTION OF LEGUMES

Most of the grasslands in our country have a very low legume component (Singh *et al.*, 1975) except those found in temperate and alpine region (Figure 30 and 31). Increase in their proportion would yield fodder superior in quality because of the higher content of protein in legumes. Legumes in general contain almost twice the crude protein content found in grasses. This difference is more pronounced in tropical regions (Anon, 1979). Legumes are also capable of fixing atmospheric nitrogen with the help of *Rhizobium* bacteria found in their root nodules. In addition to meeting their own nitrogen requirement, leguminous species improve the soil fertility also. Nitrogen fixation of 100 to 200 kg per hectare per year by white clover has been reported by Ostrowski (1972) in Australia. Most of the legumes have a longer growing season and remain green even after the common grasses have dried (Singh *et al.*, 1977; Singh

पेज 76 के बाद

Fig. 28. Regenerated grassland in low hill region in Himalayas

Fig. 29. Reseeded grassland with *Sehima nervosum* in semi arid region

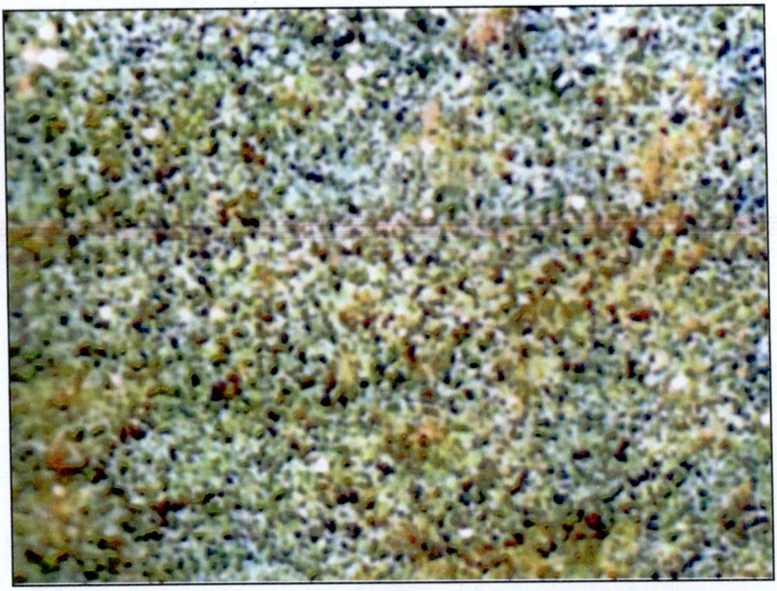

Fig. 30. A white clover dominated temperate grassland in western Himalayas

Fig. 31. Birdsfoot trefoil dominated meadow in temperate region

Table 10. Adaptability of certain forage legumes under different agroclimatic conditions

Legumes	Tropical	Sub-temperate	Temperate wet	Temperate dry	Minimum rainfall (mm)	Special Adaptability
1	2	3	4	5	6	7
Ban kulthi (*Atylosia scarabaeoides*)	+++	+	–	–	400	Drought
Calopo (*Calopogonium mucunoides*)	+++	+	–	–	1200	High rainfall
Centro (*Centrosema pubescens*)	+++	+	–	–	1200	Drought
Leucaena (*Leucaena leucocephala*)	+++	+	–	–	750	Drought gravel
Stylo (*Stylosanthes guianensis*)	+++	+	–	–	800	Low rainfall
Stylo carribean (*S. hamata*)	+++	+	–	–	600	Drought
Alyce clover (*Alysicarpus rugosus*)	+++	+	–	–	1000	Grazing
Archer dolichos (*Macrotyloma axillare*)	+++	+	–	–	1000	Grazing
Glycine (*Glycine wightii*)	+++	+	–	–	750	Drought

1	2	3	4	5	6	7
Kudzu (*Pueraria thunbergiana*)	+++	+	–	–	1250	Soil erosion
Siratro (*Macroptilium atropurpureum*)	+++	+	–	–	600	Drought
Velvet bean (*Stizolobium deeringianum*)	+++	+	–	–	1000	Meadows
Lablab (*Lablab purpureus*)	+++	+	–	–	600	Drought
Greenleaf desmodium (*Desmodium intortum*)	+	+++	–	–	900	Acidic soil
Subclover (*Trifolium subterranean*)	+	+++	–	–	1000	Meadows
Sericea lespedeza (*Lespedeza cuneata*)	–	+++	+	–	600	Poor soil
Kenya white clover (*Trifolium semipilosum*)	–	+	–	–	1000	Grazing
Birdsfoot trefoil (*Lotus corniculatus*)	–	+	+++	+	400	Grazing
Crown vetch (*Coronilla varia*)	–	–	+++	+	800	Grazing
Red clover (*Trifolium pratense*)	–	+	+++	+	700	Meadows

1	2	3	4	5	6	7
Lucerne or alfalfa (*Medicago sativa*)	–	+	+++	+++	500	Meadows
Sainfoin (*Onobrychis viciifolia*)	–	–	+++	+++	600	Meadows
Yellow sweet clover (*Melilotus officinalis*)	–	–	+++	+++	500	Meadows
White clover (*Trifolium repens*)	–	+	+++	+++	500	Grazing
White sweet clover (*Melilotus alba*)	–	–	+++	+++	500	Meadows

+++ = Good, ++ = Average, + = Below average, – = Not adapted *et al.*, 1980). Thus introduction of a suitable legume, in our grassland should be the most important aspect of their improvement (Figure 32). There are a large number of pasture legumes (Arora and Chandel, 1972; Bogdon, 1977; Heath *et al.* 1985; Skerman *et al.*, 1988) available for different agro climatic conditions (Table 10). The seed count, seed standard and seed rate of common legumes are given in Table 11 (Singh *et al.* 1998).

Table 11. Seed count, seed standard and seed rate of common legumes

Tropical Legumes

Legume	Seed count, per kg, 000	Minimum purity, %	Minimum germi-nation, %	Seed rate, kg/ha
Alyce clover	70	93	50	15
Ban kulthi	70	95	50	10
Calopo	70	94	50	3
Centro	45	94	50	5
Clitoria	30	94	50	5
Desmodium green leaf	330	95	70	2
Desmodium silver leaf	250	95	70	3
Glycine	150	98	60	4
Joint vetch	375	98	70	5
Kudzu vine	85	98	75	15
Lablab bean	4	98	60	4
Leucaena	30	98	60	4
Lotononis	3300	93	50	1
Puero	90	94	50	10
Siratro	90	98	70	3
Rabbit vine	40	98	80	6
Rhynchosia	15	97	40	6
Stylo Townsville	400	90	40	3
Velvet bean	1	99	70	30

Temperate Legumes

Legume	Seed count per kg, 000	Minimum purity, %	Minimum germi-nation, %	Seed rate, kg/ha
Alsike clover	1500	94	75	2
Birdsfoot trefoil	900	94	75	10
Common vetch	20	97	60	20
Crimson clover	320	96	81	12
Crown vetch	500	93	78	12
Greater birdsfoot trefoil	2200	94	75	5
Hairy vetch	50	97	80	30
Lespedeza	65	94	4	20
Lucerne or alfalfa	500	98	80	12
Red clover	550	97	75	12
Sainfoin	70	96	72	15
Subterranean clover	100	98	80	12
Sweet clover (white)	450	93	70	15
Sweet clover (yellow)	460	93	70	15
White clover	1750	94	75	3

Efforts should be made to maintain at least 30-40 percent of legume in a pasture mixture to obtain optimal animal production. A pronounced decline in legume percentage occurs with increase in grazing pressure. The proportion of legume shows seasonal variation too, reaching a maximum in warmer months. It also varies with phosphate and nitrogen levels, being higher at high level of phosphate application and lower at high level of nitrogen application.

Many perennial tropical herbage legumes prove to be shortlived. Persistence of four different legumes in mixture with either *Paspalum plicatulum* or *Brachiaria decumbens* under continuous grazing are shown in Figure 33 (Humphreys, 1991). The half life survival was 6 months for *Stylosanthes guianensis* cv. Graham, 15 months for *Macrotylloma axillare* cv. Archer, 18 months for *Macroptillium atropurpureum* cv. Siratro and 24 months for *Desmodium intortum* cv. Greenleaf.

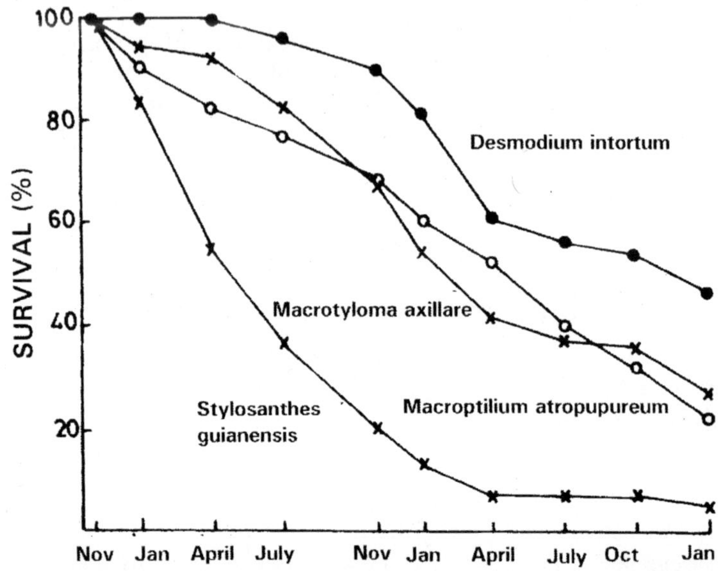

Fig. 33. Survival of four legumes on continuous grazing at Mt. cotton, Australia (Humphreys, 1991)

C. FERTILIZER APPLICATION

Grasses like other plant species require nutrients for their growth. They can not grow to their full potential unless they get essential nutrients at the optimum level. Practically none of the natural grasslands in our country receive fertilizer in any form excluding what is added in the form of dung and urine by livestock grazing them. This leads to continuous depletion of plant nutrients resulting in poor growth of forage species. As the forage yield falls, competitive capacity of desirable species also decreases resulting in their replacement by plants better adapted to lower fertility. Growth of grasses is also hindered to some extent due to lack of legumes in our grassland which are capable of fixing atmospheric nitrogen in soil through symbiosis.

The major plant nutrients for pasture species, as for other plants, are nitrogen, phosphorus, potash, sulphur, lime and magnesium. Other elements required in much smaller quantity are molybdenum, copper, zinc, magnesium, boron and iron.

Three main pools in nutrient cycle *viz.* plant, animal and soil and the main sources of inputs and losses have been shown in Figure 34 (Wilkinson & Lowrey, 1973). Ingestion of plant material by grazing animals may be augmented by supplements from off farm sources. Excretory products are added to the soil pool, urine enters the available nutrient pool or is lost to the

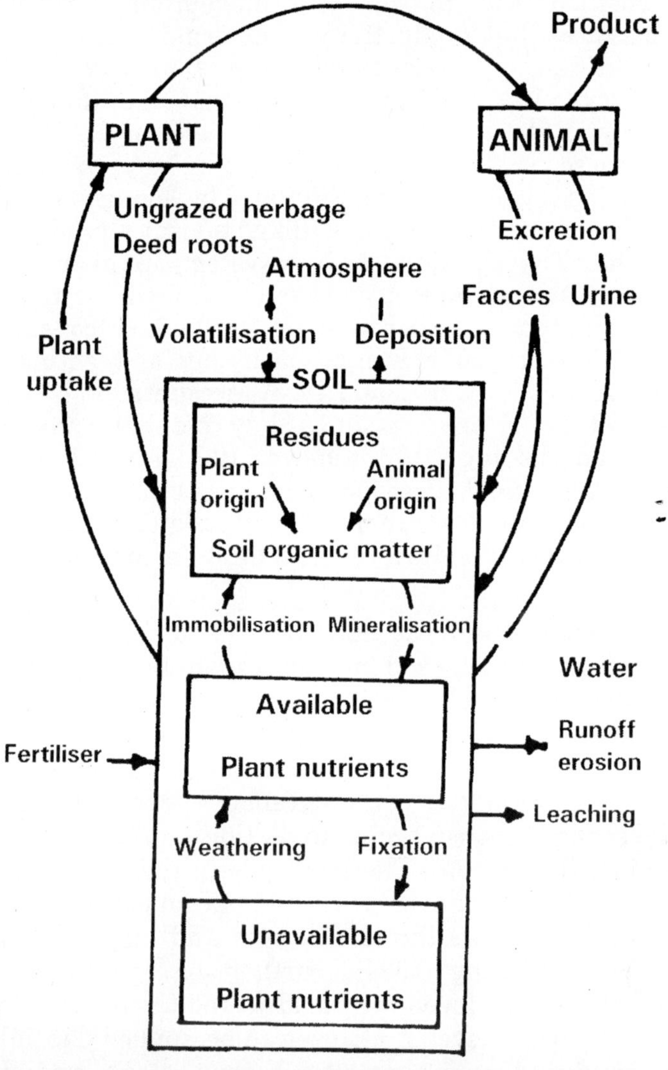

Fig. 34. Simplified nutrient cycle in pasture ecosystems (Wilkinson & Lowrey, 1973)

atmosphere whilst dung is more slowly incorporated into forms available to the plant. Nutrients are also added from atmospheric sources and from fertilizer, and are lost by leaching, erosion, volatilization and immbolization.

The size of these nutrient pools fluctuates during the year according to the balance of processes operating. A large pool size may cushion the impact of unfavourable environmental circumstances, but productivity is determined not by pool size but by the rate of nutrient and energy flow between the pools (Humphreys, 1991). Nutrients are removed by forage grasses and legumes. It is estimated that uptake of N, P and K by forage grasses is about 187, 29 and 284 kg/ha/yr, respectively, at a production level of 20 t/ha. Similarly by legumes uptake of N, P and K is about 755, 79 and 670 kg/ha/yr at a production level of 92 t/ha. The cut and remove systems represent a potent source of nutrient transfer and increased fertilizer requirement is a concomitant consideration. A 400 kg cow consuming dry matter 2.5 % of body weight each day and eating pasture containing 2% N, 0.2 % P and 1% K, accounts for 73 kg N, 7 kg P and 36 kg K in a year. A hay crop of same nutrient composition yielding 6 t/ha removes 120 kg N, 12 kg P and 60 kg K from the field. Becasue of low level of fertilizer use, the gap between nutrient removal and addition of nutrient is widening which in turn is becoming a limiting factor in production.

Nitrogen is the most important of all major nutrients in determining herbage yield but the response to nitrogen depends on an adequate supply of other nutrients and on the site class of the field. Up to a certain level of nitrogen application, the response in herbage yield is linear that is there is increase of about 20 kg dry matter per kilogram of added nitrogen. Beyond this level the response begins to decline until a maximum yield is reached (Figure 35). The point where the response drops to 10 kg dry matter of grass for each kilogram of nitrogen which is applied is defined as the target yield and target nitrogen rate (Thomas and Young, 1984). Further application of nitrogen, beyond the target levels will lead to relatively small economic gains. The target level of nitrogen to be applied depends on the grass species, growing conditions that is the site class and climatic factors. By calculating the target yield, it is possible to

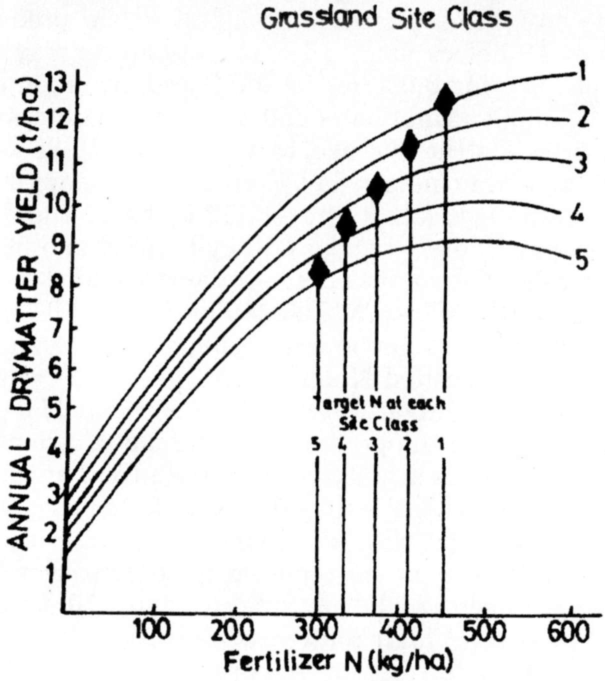

Fig. 35. Yield and response of grass according to site class, showing target, yield and target rate of fertilizer nitrogen

define an optimum yield and a practical upper limit to nitrogen input for particular conditions. Of all the elemental nitrogen applied to grassland, only 50-60 per cent is taken by plants and only 10-20 per cent actually appears in animal product, rest is lost by leaching, denitrification, volatilization etc.

The optimum dose of fertilizer for a grassland depends upon the soil type and texture, nutrient status, soil pH, herbage species and use pattern. Increase in dry matter yield of over 150% has been recorded on the application of 120 kg nitrogen and 40 kg phosphorus per hectare from grassland of moderate slope in midhills of Himachal Pradesh (Katoch *et al.*, 1975). Application of nitrogen alone was not effective in acidic soil beyond 40 kg/ha. Similarly, application of high level of phosphorus (80 kg/ha) without nitrogen fertilizer had no significant effect on forage yield.

Excreta of animals plays an important role in recycling of nutrients in grazed pasture. An adult cattle produce on an average of 12 defecations and 8 urinations per day with each defecation covering 0.093 m^2 (1.0 ft^2) and urination about 0.28 m^2 (3 ft^2). Single urination of 150 ml from a wether may occupy 0.03 m^2 area. On the average, cattle produce 20-25 kg of faeces and 8-10 litre of urine per day. Frame (1970) showed that dung spots have the equivalent of over 112 kg/ha of P and 224 kg of K. Urine spots contains about 336 kg/ha of N and 560 kg of K. Typical rates of decomposition of elements in dung pats are equivalent to 650-850 kg N, 125-400 kg P and 150-170 kg K per hectare per year (Pearson and Ison, 1987). Even though a portion of the deposited N and K may be lost by volatilization, leaching, surface run off water and other means, 50 per cent or more is recycled in the production of forage. Nutrient loss from dung spots proceeds at a slower rate than from urine spots, but up to 80 per cent of the N may dissipate from dung lying on the soil surface in warm climate. Alpine meadows in certain parts of Western Himalayas are reported to contain over 3000 kg of available nitrogen, 20 kg phosphorus and 250 kg potash per hectare because of high content of animal excreta. Excreta of animals add to fertility of soils but have adverse effect in grazing. Kemp (1984) reported reduction in herbage intake by cattle varying from 11 to 47% because of fouling of pasture.

The nitrogen cycle in a grazed grass/clover pasture is shown in Figure 36 (Walker, 1956). In addition to nitrogen that enters in cycle through fertilizer, all N which enters in the cycle is either fixed by clovers or free living organisms. Nitrogen transferred through legumes is either above ground, in the form of dung and urine, through leaching of nitrogenous compounds from leaves by rain and decay of litter or under ground through sloughing off and decay of nodules and root tissue. White clover roots and nodules contain 1.5 and 6.5% N, respectively. Small amount of direct excretion of nitrogen compounds from legume root system has also been reported.

Phosphorus requirement of grassland is high particularly for legumes which are more demanding of phosphorus. Soils in higher rainfall areas show more deficiency of phosphorus and higher dose of superphosphate (50-60 kg/ha) is required for better pasture establishment. Phosphate application usually improves the protein content of the legume.

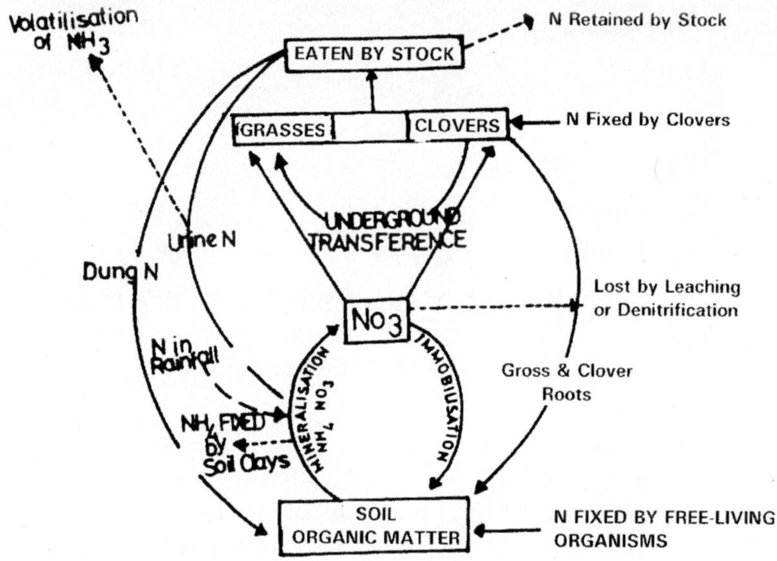

Fig. 36. Nitrogen cycle in grass/clover grassland (Walker, 1956)

Soils in high rainfall areas are invariably acid and may also be deficient in calcium and sulphur. Both of these elements are found in superphosphate and if applied in optimum dose, there may be adequate supply of these elements. However, if some areas need additional sulphur, gypsum (calcium sulphate) or sulphur fortified superphosphate (S.F. Super phosphate) may be applied for maximum pasture production. On very acid soils (pH 5.5 or less), it may be necessary to use lime (calcium carbonate). Drilling of 500 kg of lime per hectare is sufficient to enable plants particularly legumes to become established.

Potash, the third major plant nutrient is less likely to be deficient unless the pasture has been cut frequently for hay. High rainfall areas and sandy soils may show deficiency of potassium. Although optimum dose of potash fertilizer would depend upon soil test, application of 100 kg of potassium fertilizer (muriate of potash) could be desirable in potash deficient soils.

Following starter dose of fertilizer has been recommended for sown pasture (Humphrey, 1969) :

Fertilizer	Quantity/ha
Superphosphate	6 q
Lime	6 q
Muriate of potash	1 q
Copper sulphate	7 kg
Zinc sulphate	7 kg
Molybdenum	0.28 kg

Maintenance dressing of super phosphate, one quintal and muriate of potash, half a quintal are usually broadcast annually just before the main growing season commences.

Micronutrients

In addition to nitrogen, phosphorus and potash, there are a number of elements which though required in small quantity are essential for the growth of grasses and legumes. Important among them are magnesium, molybdenum, copper, zinc, manganese, cobalt, boron and iron. Rectification of their deficiencies is vital for getting optimum production for grassland. Copper and zinc deficiencies are common in sandy soils. Application of copper sulphate and zinc sulphate both at a rate of 6-8 kg/ha is recommended.

Deficiency of molybdenum interferes with the efficient nitrogen fixation of rhizobial bacteria associated with the roots of legumes. Its deficiency is more prevalent in high rainfall areas and in acidic soils. Application of molybdenum superphosphates (0.02% Mo) is commonly recommended for pastures in high rainfall area. Small dose of molybdenum 30-100 g/ha corrects its deficiency easily. Correction of molybdenum deficiency is essential for successful establishment of legumes. If molybdenum trioxide is added to superphosphate during its manufacture, it takes care of molybdenum requirement of plant.

Time of fertilizer application

In general, a constant supply of nitrogen is needed to regulate the nitrogen content of warm season grasses at all

पेज 88 के बाद

Fig. 32. Siratro established in a native subtropical grassland

Fig. 37. Temperate grassland infested with weeds

Fig. 38. Thistle - an abnoxious weed of temperate pasture

Fig. 39. Rumex infested temperate grassland in Himalayas

stages of plant growth. Split application is desirable but its value depend on source and quantity of nitrogen used, number and time of top dressing, frequency of cutting, species characteristics and climatic conditions. Grasses respond more strongly to applied N during season of rapid plant growth. Higher forage yield is produced per kilogram of nitrogen as the length of the cutting interval is extended. T-sum method is used in temperate region to determine date for first application of nitrogen. In this, mean daily air temperature from January 1 is added together and when the sum reaches 200, the time is considered to be optimum for the first application of nitrogen to grassland. The suitable time of phosphorus and potash application is the dormant season in case of perennial pasture. These should be applied as basal dose before or at the time of sowing in sown pasture. Maintenance supply should be provided on an annual basis for satisfactory growth of forage grasses and legumes.

D. CONTROL OF SHRUBS AND WEEDS

In uncared pastures, undesirable plants start appearing in large numbers. These plants compete with forage species growing in pasture for soil moisture and nutrients and affect herbage production adversely (Brockman, 1985; Williams, 1984, Vallentine, 1989; Dowling and McKenzie, 1993). Common undesirable species of temperate pastures are as follows (Figures 37, 38 and 39). Common sorrel (*Rumex acctosa*), docks (*Rumex crispers*), thistles (*Cirsium vulgare, C palustre, Carduus nutans, C. tenuiflorus* etc.), knot grass (*Polygonum aviculare*), red shank (*Polygonum persicaria*), hemp (*Cannabis sativa)*, sea warmwood (*Artemisia maritima*), fumitary (*Fumaria officinalis*), bracken (*Pteridum aquilinum*), common rush (*Juncus affusus*), matgrass (*Nardus atricata*), Moor grass (*Molinia* sp.), Heather (*Calluna vulgaris), Verbascum thapsus, Cynanchum vincetoxicum, Viburnum* sp., *Vaccinium* sp. *Polytrichum* sp., *Stipa* sp., *Sambucus* sp. poisonous peach (*Trema tomentosa*), burrow weed (*Haplopappus tenuisectus)*, goose berry *(Ribes orientale* and *Ribes glaciale*), Russian thistle (*Salsola kali)*, purple flower thistle (*Onopordum acanthium*), charlock (*Sinapis arvensis*), chickweed (*Stellaria media*), *Sonchus oleraceus, Viola arvensis, Senecio vulgaris,*

Ranunculus repens, Veronica spp., *Arnebia* sp., *Chenopodium polyspernum,* honey suckle (*Lonicera quinqualocularis and L. glauca*).

Undesirable shrubs and weeds formed in tropical and subtropical region are as follows (Figure 40 and 41); Jharber (*Ziziphus nummularia*), carrot weed (*Parthenium parthenifolium*), ratala (*Plectranthus rugosus*), Curry-leaftree (*Murraya koenigii*), prickly ash (*Zanthoxylum alatum*), wild raspberry (*Rubus niveus*), lantana (*Lantana camera)*, eupatorium (*Eupatorium adenophorum*), ageratum (*Ageratum houstonianum*), giant reed (*Arundo donax*), reed grass (*Phragmites karka*), cactus (*Euphorbia neriifolia*), Basuti (*Adhatoda vasica*), horseweed (*Erigeron canadensis*), beggar ticks (Bidens pilosa), Garna (*Carissa spinarum*), anaphilis (*Anaphilis arneosa*), Satyanasi (*Argemone mexicana*), prickly pear (*Opuntia dillenii*), Mainphal (*Randia dumetorum*), ban (*Vitex negundo*), barberry (*Berberis aristata*), kana (*Saccharum munja)*, Manidhar (*Dodonia viscosa*), common thorn apple (*Datura stromonium*), Noogorra burr (*Xanthium pungens*).

Plants such as *Aconitum heterophyllum, Adonis aestivalis, A. chrysocyathus, Aquilegia vulgaris, Anemone obtusiloba, Actinea richardsonii, Cinicifuga foetida, Ranunculus lactus, Oxytropis lambertii, Zigadenus gramineus*, catsear *(Hypochoeris* spp.), fireweed (*Senecio lautus*), crown beard (*Verbesina encelioides)*, rattle pod (*Crotalaria* spp), selanium weed (*Neptunia amplexicaulis*), mintweed (*Salvia reflexa)*, gomphrena (*Gomphrena celosioides)*, blue heliotrops (*Heliotropium amplexicaule*), mallow (*Malva parviflora*), caltrop (*Tribulus terrestris*) and *Stipa sibirica* are reported toxic to livestock (Stefferud, 1948; Chopra *et al*. 1965). Their eradication from pasture is also necessary for getting good quality forage. These noxious plants can be removed by mechanical or chemical control measures (Figure 42 and 43). Herbicides like gramoxone, dalapon, 2,4-D, glyphosate or picloram could be used to check growth of weeds (Singh, 1988).

Mechanical plant control

The mechanical plant control includes hand chopping or cutting, girdling top pruning, grubbing, bulldozing, blading, chaining, cabling, railing, roller chopping or crushing,

Fig. 40. Cactus - an undesirable shrub in grassland of tropics

Fig. 41. Lantana - an abnoxious shrub infesting grassland in tropics and sub tropics

Fig. 42. Removal of shrubs leads to regeneration of grasses

Fig. 43. Hill grassland under regeneration after removal of undesirable shrubs; Hybrid Napier plant in hedge row.

shredding, mowing, harrowing, ploughing, offset-disking, raking etc. Selection of suitable mechanical method keeping in view the site topography and terrain, area to be covered, nature and density of noxious species, site potential management objective, follow-up required and resources available is important. Hand chopping or cutting has disadvantages of being slow and adapted to small area. It helps in maintenance of cleared areas against reinvasion or for removal of limited stands of poisonous plant. Girdling can be done by hand saw and is confined to kill weed trees. Pruning is done to rejuvenate woody plants through stimulation of regrowth while removing overstory. Grubbing is used for entire plant removal and is adapted to scattered herbaceous plants or scattered shrubs or tree seedlings of small diameter. It involves digging out plant with its root system. Hand grubbing is slow, labour intensive and can be used only to small areas of sparse stands. Ploughing, disking or harrowing by different means can be done for removal of all plant species and preparing seed bed.

Herbicidal plant control

Herbicides (weedicides) are chemicals that kill plants or severely interrupts their normal growth process. They are an effective means for the control of noxious plants in grasslands. Selection of herbicides, their concentration, application technique and time of application is important for their efficient use. Herbicidal control method has a number of advantages over other control measures. They can be used easily on steep, rocky, muddy areas, provide rapid control have low labour requirement and can be selectively applied with minimum damage to desirable plant species.

Herbicides are of following types (Vallentine, 1989) :

(i) Contact herbicide : A herbicide that causes localized injury and only to plant parts directly contacted.

(ii) Systemic or Translocated herbicide : A herbicide applied to one part of a plant but which is spread throughout the plant where effects are produced.

(iii) Selective herbicide : A herbicide that kills or damages a particular plant species or group of species with little or no injury to other plant species.

(iv) Non-selective herbicide : A herbicide that generally kills or damages all plant species to which applied.

(v) Soil-active herbicide (Soil sterilant) : A herbicide which kills or damages plant when present in soil. The effect may be temporary or permanent and selective or non selective.

The common properties of herbicides used in grasslands are given in Table 12.

Table 12. Properties of herbicides used in grassland

Herbicide	Group and type of herbicide	Uses
2,4 – D	Phenoxy, selective, translocated foliage	Effective on many broad-leaved herbaceous plants and some shrubs
Glyphosate (Round up)	Aliphatic, non selective, translocated foliage, broad spectrum	Used in total plant control, often applied with 2,4-D Persists 1-3 weeks in soil
Paraquat (Gramoxone)	Bipyridyl, selective to non selective, contact, foliage	Used as spot treatment, can be applied just prior to range seeding, has minor effect on broad leaf perennials
Picloram (Tordon)	Picotinic, selective, translocated, foliage or soil, non volatile.	Effective as foliage spray or pellets on leafy spurge, prickly pear, thistle etc
Triclopyr (Garlon)	Phenoxy-picolinic, selective, translocated foliage applied	Effective on broad leaf weeds and shrubs, also effective in basal spray and trunk injection Degraded rapidly in soil, can be mixed with 2,4-D
Atrazine	Triazine, selective, soil active herbicide	Kills annual grasses and broad leaf weed seedling, providing chemical fallow or range Persists for over 1 yr in dry soil
Dalapon (Dowpon)	Aliphatic, translocated selective foliage	Foliage spray on emerged aquatic such as cattails and rushes, persists in soil up to 2-6 weeks

Amitrole (Weedazole and amino triazole)	Trizole, foliage, non selective, translocated	Effective on thistle horsetail, cattails poison ivy, suggested for patch treatment, persists 2-4 weeks in soil
Kerosene or Diesel oil	Organic compound non-selective, foliage or trunk applied	Mostly used in basal trunk spraying, minimum persistence in soil

Method of application

(i) Foliage application : Broadcast spraying is the application method most commonly used on grassland. When herbicide is applied on all plants, desirable or undesirable, selective herbicides are generally required. The optimum application rate for broadcast spraying herbicides is determined by formulation, environmental, climatic, plant and other factors interacting together. With aerial application, spray volume can be decided, keeping in view foliage and spraying machine.

(ii) Stem application : The treatement of trunks or stems of individual woody plants is generally practical only where , there is a sparse stand of single stem trees or where small areas of grassland are to be treated. There are five different methods of herbicide application to woody plants.

(a) Trunk base spray.

(b) Placement in frills , notches or girdles. Frills are made by overlapping downward axe cuts through the bark around tree trunk near the ground line.

(c) Streamline application in which a 4 to 6 inch strip is sprayed on juvenile bark.

(d) Cut-stump treatment.

(e) Trunk injection. Tools for tree injections have a hole in a chisel point through which herbicide is injected after the chisel is driven into the stem. Injections are commonly made in cuts about 5 cm apart at the base of the stem.

Picloram, triclopyr and hexazinone are highly effective in individual trunk treatment. 2,4-D is also effective on many woody plant species. Trunk base spray is applied to the lower 12 to 18 inches of trees trunk until it runs down the bark channels to the ground line. This method is effective against tree up to 12 cm in diameter. Notching or girdling before basal spraying may be needed on larger stems or stems with thick bark.

(iii) Soil Application : Soil application of herbicide is another means of killing undesirable plants. Soil active herbicides are more commonly applied as dry granules or pellets than as spray since vegetation intercepts some or most of the spray. It is less dependent on stage of plant growth than foliage sprays but does require rainfall for herbicide to dissolve and penetrate into the soil.

E. GRAZING MANAGEMENT

The productivity of animal may be optimal at low stocking rates but this may result in low utilization of the pasture. At moderate stocking rates, production per animal may decrease but production per unit area increases.

At high stocking rates production per animal as well pasture decline as animal demands exceed the supply (Lazenby and Swain, 1972).

1. Grazing System

There are different systems of grazing like controlled, continuous, deferred, rotational, and periodic grazing. Each of these systems has its merits and demerits. However, control of the number of animals grazing under any system is of prime importance. This should be determined on the basis of the carrying capacity of the grassland. Carrying capacity of a grassland is defined as the number of animals that can graze in an unit area without over grazing or under-grazing in an average season. Grazing system in practice are as follows:

a. *Continuous grazing or set stocking*

System of grazing in which animals are allowed to graze freely over a particular area continuously is referred as continuous grazing. If number of animals allowed to graze is higher than the carrying capacity of grassland, there is fast depletion of desirable species and deterioration of grassland in due course of time. In most of the grassland in our country, grazing pressure is high and it does not give chance for reseeding which eventually leads to decrease in ground cover and degradation of grassland. If the number of animal is controlled in continuous grazing, the sward becomes denser and the proportion of desirable species remains satisfactory. Continuous grazing is prevalent in most of grasslands in our country. Because of high grazing pressure, limited area under fodder crops and lack of grassland management practices, the productivity of grassland and animal is very low. There is continuous depletion of desirable herbage species and increase in the degradation of grassland due to lack of reproduction and reseeding of grasses (Figure 44).

Fig. 44. Livestock grazing in a temperate pasture in Himalayas

b. *Rotational grazing*

In rotational grazing, grassland area is divided into two or more blocks or paddocks and grazing is allowed in rotation at suitable interval during the grazing season (Figure 45). Number of animal is maintained according to carrying capacity of grassland. A well managed grassland can sustain stocking rate of 5 animals/ hectare. Normally 1 hectare is allowed for 100 animals for each day grazing. The recovery period between grazing depends upon the agro-ecological situation and grass species. An interval of 3 to 4 week is usually adopted successfully. The change over from one compartment to another provides the advantage of giving animals a young, palatable and nutritive herbage and provides a period of rest to grasses during growing season which results in good regeneration by vegetative means. However, this system does not give chance for seed formation.

Fig. 45. Grazing paddock for rotational grazing in a grassland

c. *Deferred grazing*

This system of grazing consists of subdividing grassland into three compartment with grazing allowed alternately in two and third completely closed during the growing season until seed maturity. This gives grasses a chance to make sufficient growth and accumulate sufficient reserve food material to maintain the vigour of plants and regenerate from the seed that are shed. Grazing is deferred in different compartments each year. This way each compartment gets adequate period of rest and chance for recovery.

d. *Hohenheim system*

It is an improved system of rotational grazing developed in Germany during First World War, to produce sufficient milk without any need to feed concentrates to milch cattle. It consists of dividing the pasture into several equal sized plots or paddocks and applying large quantities of nitrogenous fertilizers in these paddocks. The dairy cattle are separated into three groups, high, medium and poor milkers and let into the paddocks in rotation, so that the best milkers will get the youngest and most nutritious grazing (Narayanan and Dabadghao, 1972).

e. *Deferred- rotational grazing*

This is a combination of deferred and rotational grazing in which grassland is divided into three compartments and grazing is allowed in rotation so that each compartment is grazed for one-third of grazing season and protected afterwards until the seeds mature (Narayanan and Dabadghao, 1972). By the time animals are allowed to graze in third compartment seed would have matured and got shed. Grazing in each compartment is rotated every year. Assuming growing period of three months, grazing plan for the three compartments can be shown as follows :

	July	August	September
First year	I	II	III
Second year	II	III	I
Third year	III	I	II

f. *Strip grazing*

This system is a form of controlled grazing system. Livestock is contained within a given area or strip and a daily or twice daily allocation of grass is made to herd. It is very efficient as a means of rationing since by reducing the area on offer, utilization can be very high. It requires careful decision each time during growing season to allocate the area to be grazed. Precaution has to be taken not to restrict maximum intake and therefore production. This system has a high labour requirement and requires electric fencing and is not applicable to average condition of grassland.

g. *Mob grazing*

This is a form of rotational grazing in that a large numbers of animals are concentrated when grazing and are moved from paddock to paddock, but it differs from true rotational grazing in that grazing is continued in any one area until the feed is eaten down evenly and closely. It is used as a method to clean up coarse fibrous and rank feed left behind as a result of previous poor utilization.

h. *Periodic grazing*

In this system the whole of the grassland area is grazed or cut for hay for a period of a year or two but completely closed in the following one or two years to enable self seeding and recovery of the stand.

i. *Zero grazing*

Zero grazing refers to old practice of soiling or feeding fresh cut crops to household livestock. The soiling includes a wide variety of arable and forage crops. Zero grazing relates to feeding of crops which are normally grazed. The main advantage of zero grazing is that it reduces wastage caused by selective grazing, trampling and fouling of pastures and therefore results in higher output per unit area. There is higher production in zero grazing but the heavy labour demand makes it too expensive. Zero grazing is more suitable for tall forage species where there can be excessive damage by trampling and fouling of crop used *in situ.* Zero grazing is suitable at farms where land holding is small and grassland is easily approachable and not too far away from the livestock shed.

Generally uncontrolled continuous and over-grazing is prevalent in most of the grasslands of our country. This results in elimination of desirable species and in turn loss of standard predominance of species which are not so palatable and nutritive. Though the rotational system of grazing may give superior result, it is not so easy to adopt on a large scale under existing conditions. The most desirable system of grazing, thus would be that of periodic grazing or deferred grazing limiting the number of livestock on the basis of the carrying capacity of grassland. These systems would give a chance for self seeding and stand recovery.

Most of the available herbage in grassland in our country is utilized by grazing excluding that found in meadows around villages. Meadows are harvested during August and September for hay making which is used for feeding during winter. There are several pastoral communities, located in northern parts of the country covering states of Jammu & Kashmir, Himachal Pradesh and Uttar Pradesh. Keeping in view the practice followed by these pastoral communities, the system of grazing may be categorized either on the basis of methods of grazing or patterns of migrations (Tucker, 1986; Singh, 1988 and Singh, 1992). As regards method of grazing, mostly the flock herds move from one area to another when grazing become thinner. On the basis of migration, four grazing systems have been recognized which are as follows :

(i) Total nomadism : In this system, animal owners do not have a permanent place of residence and they keep on moving alongwith families and herds in search of forage. Some of the Gujjars and Bakarwal of Kashmir valley,

Gaddis of Chamba and Kangra in Himachal Pradesh and Bhots and Khadwals of Uttarakhand are communities practising total nomadisms.

(ii) Seminomadism : It is a related system when animal owners have a permanent place of residence near which supplementary cultivation is practised, but for greater part of the year, they travel with their herds to a distant grazing area. Kinnauris of Himachal Pradesh and some Gujjars of H.P., Punjab and Haryana practise seminomadism.

(iii) Transhumance : This system is practised by farmers with a permanent place of residence whose herds are tended by a herdsman for long periods of time in the distant grazing areas. Chaupan Shepherds of Kashmir belong to this category. They do not have their own animals but graze the flocks of Bakarwal and Gujjars.

(iv) Partial nomadism : Farmers who live in permanent settlement and keep herds in the vicinity belong to this category. Some of the Gujjars of H.P., J & K and U.P., and Todas of Nilgiris in Tamil Nadu practise partial nomadism. Village grazing system may also be included in this system.

2. Stocking Rate

Stocking rate is the actual number of animal or animals unit on a specific area for a specific period of time, usually for a grazing season. It is also defined as the number of animals grazing a unit area (usually 1 ha) at a particular time.

The stocking rate determines levels of animal performance. Animal compete for available forage in pasture. At low stocking rate, animals have opportunity for selection and their productivity is high but as stocking rate increases, the opportunity for selectivity decreases, and the reduced nutritive value of diet ingested decreases individual animal performace. Simplest diagrammatic relationship (Figure 46) between stocking rate (beasts ha^{-1}) and animal output per head and per unit area under tropical condition has been shown by Jones and Sandland (1974).

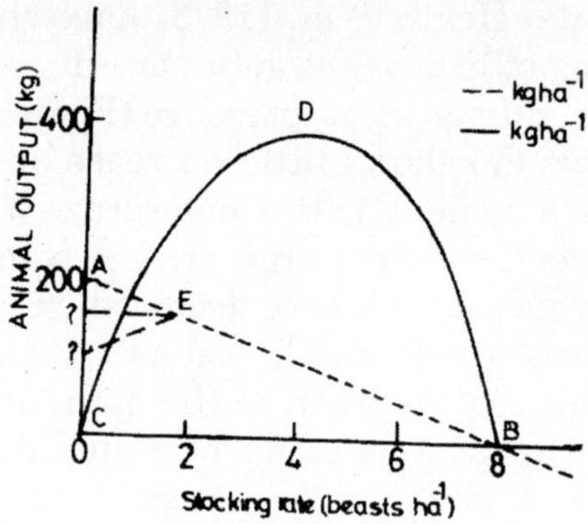

Fig. 46. Diagrammatic relationship between stocking rate and animal out put (Jones and Sandland, 1974)

They found that linear relation between gain per animal and stocking rate more closely fit a linear model than the curvilinear model. The linear model predicts that gain per animal at the optimum rate will be one half that at the lower rate. Negative value for live weight gain per animal would occur at a stocking rate double of the optimum rate. The gain per hectare would change gradually on both sides of the optimum rate and their would be no sharp declines. Stocking rate should depend on rate of growth of forage species, fertilization scheme, nutritive value of herbage, botanical composition and ground cover, accessibilty of forage, class of animals and nature of animal product.

In "put and take" system of grazing, stocking rate is intentionally varied according to available herbage in the grazing season, so as to achieve a constant grazing pressure.

The method involves two classes of animals :

(a) Testers : Those selected to remain on the experimental plot for the duration of the full grazing period and about which inference are to be made about herbage quality by average daily animal performance.

(b) Put and take animals (also called regulators or grazers) : That are placed on paddock where available herbage exceeds the daily requirement of testers and removed when quantity of herbage declines.

There is strong relationship that exists between production

and stocking rate. Heath *et al.* (1973) have shown relationship in relative terms of liveweight gain per animal and per hectare based on an optimum grazing pressure (Figure 47). The product per animal curve (Y) shows little decrease as stocking rate (X) increases from a lenient rate (undergrazed) to the optimum rate. There exists a rather narrow critical point at the optimum rate beyond which a marked decrease occurs. Product per hectare (Z) shows a very rapid rise as stocking rate increases from an undergrazed condition to the optimum rate. Maximum output takes place at a stocking rate slightly in excess of the optimum, then a sharp drop occurs. Higher stocking rate results in damage to grass cover due to treading as there is mechanical damage to pasture plants and soil by hooves. The pressure exerted on a grassland by sheep is estimated to be 0.8 - 0.95 kg per cm^2, that exerted by cattle to be 1.2 - 1.6 kg per cm^2.

It is estimated that grazing animals tread on 0.01 ha per day although, the actual area affected depends on the availability of feed, animal behaviour and the weather. On treading, tillers and shoots are torn from the crown, stems and leaves are bruised. Similar damage may occur to the roots if the soil is very wet and pliable. Destruction of soil structure with a resulting loss of permeability to gases and water also causes poor plant growth. These all results in decrease in pasture yield. Studies on effect of treading on pasture yield have shown yield of 88, 68 and 34 per cent of not trodden pasture yield at 10, 25 and 50 ewe equivalent stocking per hectare, respectively.

Average stocking rates on permanent grassland range from 0.1 animal unit per hectare in Australia to 1.5 animal units per hectare in Europe and 9 animal units per hectare in parts of Asia. These averages mask the variation in stocking rates which occurs within each country (Christie, 1984).

Pasture vigour and quality on stock camps may be very high and total pasture production much greater than on slopes. On large unimproved holdings, an extensive set stocking system still prevails (1 sheep/1-2 ha) with traditional ewe, wether and hogget blocks. Under intensive hill farming, high carrying capacity (12-15 sheep/ha) may occur.

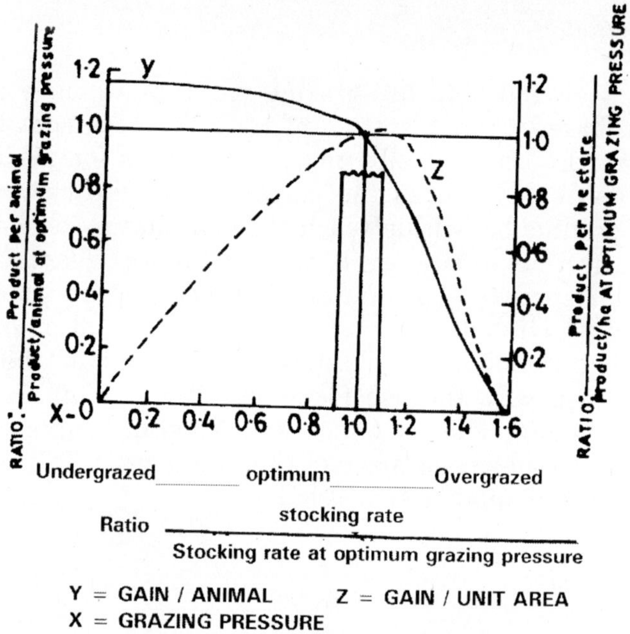

Fig. 47. Relationship of product per animal and per hectare to stocking rate (Heath *et al.*, 1973)

In hill country pasture in New Zealand, wool produce per hectare increased with stocking rate up to 16.1 ewes/ha then started to decline. Generalization of this behaviour is common to many stocking rate and productivity experiments (Mott, 1960; Heath *et al.*, 1973).

Grazing Pressure

The term grazing pressure is used to relate the number of animal units to the amount of forage availability. Mott (1960) defined grazing pressure as the ratio of feed demand to feed supply.

$$\text{G.P.} = \frac{\text{D. M. demand animal}^{-1}\ \text{day}^{-1} \times \text{No. of animal ha}^{-1}}{\text{D.M. available day}^{-1}\ \text{ha}^{-1}}$$

It will be light during the season when supply due to growth increment exceed the demand and vice versa.

The use factor

It is an index to the grazing use that is made of forage species, based upon a system of range management that will maintain the economically important species for an indefinite time. It is expressed as the percentage of the current year's weight production, within reach of stock, that is consumed. The factor affecting use factor is the preference of animals for a particular species. A use factor of 50 per cent indicates that a plant will have half of its total available annual production of vegetation removed by livestock at the end of the grazing season. In a way, the use factor is a measure of the relative value of a plant. The use factor is the total utilization of species after full use has been made of the range and it is a product of preference and quality available.

Forage density

It is referred to only that vegetation which is within reach of livestock. Vegetation not available to stock is excluded from the estimate of forage density.

Forage acre

It is theoretical acre totally covered with vegetation, all of which is properly usable by livestock.

Forage acre = forage acre factor x surface area

Forage acre factor is an expression of relative forage value. It is computed as follows:

Forage acre factor = Average density of range forage x proper use factor (Average weighted palatability)

Thus an acre only half covered with vegetation (density = 0.5) of which only half is properly edible to stock (use factor = 0.5) would have a forage index or forage acre factor of 0.25 or the equivalent of 1/4 forage acre.

3. Fencing

Fencing in general helps in regulating grazing and also in protecting a particular grassland area. Barbed wire, dry wall (stone), electric wire or live hedge can be used as fence to control or prevent grazing in grassland. Dry wall, barbed wire and live hedge fencing are used mostly on boundaries as demarcation of holding and for exclusion of stray animal. Electric wire fencing is used to regulate grazing under advanced grassland management. It is common in advanced countries.

On enclosure of a grassland area, stocking becomes more intensive and there is uniform grazing of all herbage available. After removal of animal, there is also uniform regrowth. Fencing of grassland without proper grazing management may lead to appearance of undesirable herbaceous weeds and shrubs. The barbed wire fence is the most common type of fence. It consists of strands of barbed wire laid horizontally on wooden, cemented or angle iron posts. Normally 3 or 4 stranded barbed wires are used. The poles are usually of 2 m height and have diameter of about 20 cm. The lower portion of poles are dipped in coaltar to protect against termite and rotting. About 24 poles are required for fencing each acre of land with spacing of 5 m between poles. The spacing of strands above ground level may vary. Strands at 15, 45, 83 and 128 cm above ground is quite effective. A spool of barbed wire weighing about 40 kg may have wire length of about 175 m. In line or portable stiles are used on fence for crossing over.

Stone wall used for fencing should be 60 to 75 cm wide at the base and 45 to 60 cm wide at the top. The height of the wall is usually kept between 1 m to 1.5 m.

Fencing by mechanical means is an expensive proposition due to high cost of material and labour. However, use of live fence particularly on boundaries of grassland could be desirable. Suitable species for live fence under different agroclimatic situations are listed below :

Tropical and sub tropical region

Bamboo	(*Dendrocalamus strictus*)
Barberry	(*Berberis aristata*)
Century plant	(*Agave americana*)
Chaste tree	(*Vitex negundo*)
Chhun cactus	(*Euphorbia royleana*)
Golden dew drop	(*Duranta phemieri*)
Jharber	(*Ziziphus nummularia*)
Lantana	(*Lantana camera*)
Natal plum	(*Carissa spinarum*)
Physic nut	(*Jatropha curcas*)
Prickly ash	(*Zanthoxylum armatum*)
Prickly pear	(*Opuntia dillenii*)
Sorrel	(*Dodonia viscosa*)
Thatch grass	(*Saccharum spontaneum*)
Vetiver	(*Vetiveria zizanioides*)
Yellow raspberry	(*Rubus ellipticus*)

Temperate region

Barberry	(*Berberis aristata*)
Big sagebrush	(*Artemisia tridentata*)
Currant	(*Ribes sanguineum*)
Lantana	(*Lantana camera*)
Myrtle	(*Lagerstroemia indica*)
Plectranthus	(*Plectranthus rugosus*)
Oak	(*Quercus incana*)
Yellow raspberry	(*Rubus ellipticus*)
Black locust	(*Robinia pseudoacacia*)

F. SOIL AND WATER CONSERVATION

Most of our grasslands particulary in tropical and subtropical region are located on poor soil deficient in plant nutrients. Over grazing is causing loss of vegetative cover which is in turn is resulting in their further degradation. Treading increases mortality and reduces life span of seedling. It causes packing of the soil particles and a loss of the larger pores in the soil mass results in an increase in bulk density. The damage resulting from the mechanical disturbance of a wet soil known as poaching, pugging or pudding arises when the bearing strength of the soil and vegetation cover is exceeded. These changes in total porosity reduces aeration, moisture infiltration, moisture retention and drainage. Due to lack of adoption of any soil and water conservation practices, there is high degree of soil erosion resulting in loss of plant nutrients. Erratic rainfall distribution further aggravates the problem of soil erosion and limited growing season.

On depletion of vegetative cover, the erosion starts which damages even existing vegetation leading to conversion of grassland into degradation which further results in formation of barren land. The basic principle of soil and water conservation is to prevent or reduce the loss of soil and moisture either by vegetative or mechanical means. Prevention of soil loss by vegetative mean could be achieved by reducing grazing pressure, reseeding suitable grass legumes mixture, or closing grassland to grazing allowing regeneration of native flora. Planting of suitable grasses like *Vetiver, Saccharum*, and *Eulaliopsis* hedgerows on contour is also useful in soil and water conservation.

Though vegetative measures are most economical and convenient, mechanical measures are also required to be adopted under certain situations. Some of the mechanical measures effective for soil conservation are contour furrowing, contour trenching and control of gullies.

Contour furrows

Contour furrows are small level channels or excavations constructed on contours to store water and allowed to be held inside the ground level. The excavated material is put on the down stream side (Lal, 1990). These furrows can be constructed by using hand plough or tractor drawn disc plough. It is desirable to leave a strip of undisturbed soil between the furrows. The furrows are normally 10 to 20 cm wide, 10 to 15 cm deep and 1.0 to 2.0 m apart. These furrows make use of additional water and regrass more rapidly. It is suitable on sites having slope up to 15 per cent and is not recommended on steep slopes. It has not shown good results on shallow and unproductive soil.

Contour trenching

This involves excavating trenches along the contour or across the slope of the land, generally in the top portions of the catchments. These trenches conserve and spread a portion of the run off and allow it to percolate into the soil slowly. With the retention of water and checking of the run off, there is improvement in vegetative cover and growth and reduction in soil erosion.

Generally trenches are 30 to 60 cm wide, 30 cm deep and 5 to 15 m long depending upon site. Trenches should run perfectly level so as not to allow the trenches getting converted into gullies. In case of continuous trenches which are running perfectly level, interruptions or equalisers at every 5 or 10 m should be provided so that these may not turn into water channel. The soil excavated is used to form a band on the down stream side leaving a berm equal to the depth of the trench. Seeding of grasses can be done in patches where silt have been collected on the bunds or on the berms. Trenches are effective in reducing the velocity of run off and controlling soil loss, water storage for longer time on slope and is utilized by plants for growth.

Control of Gullies

If the rill or sheet erosion continues, it leads to formation of gullies. It is caused in loose textured soil lacking vegetative cover. It starts in natural depression in overgrazed area, livestock trails etc. Intense rainfall is also responsible for gully formation. Following measure could be adopted for control of gully formation :

(a) Closure of area to grazing : Prevention of grazing helps in rejuvenation of vegetative cover. Grasses get chance to seed and to spread by vegetative means. On the improvement of vegetative cover, soils are bound with root system of plants preventing erosion and reducing run off.

(b) Construction of contour and peripheral bunds : The construction of bunds prevents water from coming into gullies and increases the moisture retention capacity of the soil. These bunds are stabilized by sodding with grasses. The excess runoff can be dispersed off safely by digging diversion channels above gully head.

(c) Easing of the head of the gully : This is done by partly cutting at the top and filling it at the base. The head can be sloped to give an angle of repose required by the soil. The slope can be stabilized by planting grasses which spread rapidly.

(d) Gully plugging : It is done to check the upstream progress of gullies by reducing the grade at the top of the slope. Brushwood, live hedges, earth, sand bags and boulders can be used for gully plugging. It is quite effective in retarding the run off, increasing the percolation and silting and thus improving the soil moisture regime (Figure 48).

Pitting

The formation of small basins or pits or other indentations of the soil surface is called pitting. It is quite an effective mechanical measure to reduce run off in low rainfall areas. In heavy rainfall areas, shallow pits are made. The moisture so conserved are used for plants for their growth. It is not recommended in sandy soils or where there is very little run off. It may thin out the vegetation but provides additional moisture for remaining plants for longer period of growth.

Ripping

It includes subsoiling or deep chiseling to break compacted layers in the soil that inhibit root development and moisture

Fig. 48. Stablization of gully through a series of stone check dams

penetration. It is applicable only in grassland which are more or less level or having undulating topography.

Basin listing

Making the small interrupted basins along the contour is called basin listing. It is suitable in low rainfall areas and in areas with gentle slopes. Listing or ridging are terms used to describe the formation of alternate furrows and ridges on the land. The construction of small dams at intervals in the furrows giving the appearance of basins is also known as basin listing. It helps in retention of water which promotes plant growth. The practice has not been found so successful in areas with high rainfall and soil with poor permeability as it causes water logging and there may be damage to ridges by over topping.

Water spreading

The process of diversion of water from natural water source to adjacent low lying and flat areas is called water spreading. It is done by constructing dykes and dams. To promote vegetative cover in grassland and rangeland, for preventing soil erosion and degradation, additional supply of water particularly during moisture stress is required to be made.

Water spreading is an effective measure to check soil erosion losses and damage caused by floods. It is suitable in low rainfall areas and is practiced in areas with gentle slope and with uniform surface features. It is more effective if soil has good permeability. However, loose sandy soil or rough broken terrain are not suitable for water spreading. There should be even spreading of water in all areas and it should not exceed the optimum requirement of the area. A small dam placed in a water course with gradient ditches or terraces leading the water out to gentle slopes where it is released is the most common type of water spreader. If terraces or ridges are used for diversion, these should be provided with frequent openings at several places to regulate the amount of water for water spreading (Lal, 1990).

Small graded furrows, ditches or ridges can be used for water spreading where run off water is concentrating in a broad, shallow gullies. On uniform slopes, contour furrows and ridges can be more suitable in minimizing concentration of water in natural depressions and to increase absorption. A combination of earth fill dam and broad base earth dyke is generally used wherever possible to turn all the water out of active gullies (Figure 49).

Water should be released only at points where it will spread on the land as evenly as possible, and not tend to concentrate in depression and cause gullying.

Irrigation

It involves providing of water supply to areas having a deficit of soil moisture. Because of irregular rainfall, growth of grasses in grassland is limited to 4-5 months in a year. In high altitude dry temperate zone in Ladakh region of Kashmir, Pangi, Lahaul Spiti and Kinnaur in Himachal and part of Uttarkashi, Chamoli and Pithoragarh in U.P. Hills, it is common pratice to irrigate grassland and meadows in the vicinity of settlements. These meadows are irrigated by flood irrigation diverting snow melt water comming in open channels from higher elevations.

Fig. 49. Water harvesting through earthen dam-improving water status and providing water for irrigation and livestock

Due to dry weather which prevails in dry temperate region because of very low rainfall during summer, even grasses do not show satisfactory growth and irrigation becomes essential to obtain higher productivity of herbage. Meadows in this region are provided 3 to 5 irrigation, during June to August depending upon water availability and rainfall.

Surface or sprinkler irrigation could be provided in any grassland for increasing productivity and period of green herbage supply. Irrigation of grassland depends upon availability of water and condition of grassland. In general, it is confined to only government dairy farms or other artificially established grassland with good soil fertility and higher production potential.

G. DRAINAGE

Lack of proper drainage results in waterlogged soil which promotes growth of undesirable plant species such as rushes, sedges, phragmites, cattails and polygonum. There is not only dominance of unpalatable and undesirable species, restriction is imposed on grazing also. There is severe poaching of land by the treading of stock. Wet land has also other drawbacks such as harbouring of parasitic organisms harmful to livestock, difficulty in harvesting of herbage and adoption of any grassland improvement practices like seeding of grass or legume and application of fertilizer. Sometimes water logging is due to presence of iron pan which holds up water near the surface and so prevents free percolation. The pan is required to be broken for proper drainage. In many cases deep ploughing is the simplest way of tackling the problem.

On the boulder clay soils and where alternating beds of impervious clay and porous rock exist, springs of water are a source of wet patches. Drainage problem is also caused when a porous sand stone overlies an impervious clay or shale sub soil. These areas can be improved by providing proper drain. Surface wetness is also due to presence of surface "mat" of undecayed herbage. The peaty layer acts as a sponge and soaks up the rain water holding it near the surface and rendering conditions favourable for the growth of sedges, rushes etc. where the grassland is improved by breaking up the mat and applying lime, the surface in many cases dries up and the moisture loving weeds disappear as the stock graze more closely on the improved herbage (Moore, 1950). Wetlands could be drained completely merely by cleaning out over ground ditches and providing proper drainage based on slope and depth. Proper cleaning of water courses and providing desirable gradient is helpful in improving wetlands to make them productive for livestock grazing.

H. BURNING OF GRASSLAND

Burning is a convenient method of removing unutilized herbage residues of previous year and to control the growth of undesirable shrubs and weeds. Burning also promotes early spring growth of grasses. It increases palatability of forages and removes old, dead material and thereby increases utilization, by grazing animals. It also increases forage production thus grazing capacity. It releases plant nutrients in the soil and litter for plant use.

Burning reduces heavy needle and debris mats that inhibit growth of grass and other desirable plants. It also prepares seed beds for natural and artificial seeding of forage species. It reduces the incidence of insect and fungal disease attack also (Vallentine, 1989).

Burning of dry vegetation in our grassland is a wide spread practice in many parts of India. But it has to be practiced with great care. In humid and high rainfall area, burning may not be harmful, but in arid and semiarid region burning is not a desirable practice. It is considered antagonistic to natural succession and may result in favouring certain species over others and may not be favourable from the grazing point of view. For example, with *Sehmia-Dichanthium* type of grassland, burning leads to a dominance of the *Cymbopogon* species which are worthless from the grazing point of view, but in the *Phragmites- Saccharum* type, burning results in bringing about a reduced form of *Imperata cylindrica* which is useful for forage purposes (Narayanan and Dabadghao, 1972). There is loss of organic nitrogen, carbonaceous material and organic matter on burning. It bares the soil to erosion and water run off deprives the soil of litter and mulch and reduces the effectiveness of rainfall.

Burning should be done when grasses are dormant for a rapid effective burn without damage to living parts. A rest period is needed before burning to allow an accumulation of combustible material. Burning every third or fourth year in a properly managed grazing system is sufficient to check bush enchroachment and provide optimal forage production. A post burn rest is needed to enable grasses to recover and build up food reserves before being grazed, but the optimal period is difficult to establish. If rainfall ensures vigorous regrowth, grazing within 4-6 weeks does not harm grass cover. Temperature at soil level on burning frequently exceeds 500°C but are less severe above 1.0 m height. At 2 cm soil depth, it may fluctuate as much as 14°C above normal but often as little as 3-4°C.

Grasslands are less susceptible to the effects of drought in the absence of burning, since previously unconsumed standing pasture is available to meet a seasonal shortfall in forage supply (Humphreys, 1991). The rapid restoration of pasture cover and provision of forage for animals are ensured if burning is restricted to occasions when adequate soil moisture is present.

5

FODDER TREES AND BROWSE

Fodder trees and shrubs provide foliage when native grasses in grassland have dried or are dormant. The plantation of fodder trees at wide spacing (25m x 10m) in and around grassland areas and waste and barren land could be extremely advantageous. In addition to supply of green fodder during scarcity periods, their planting at wider spacing do not hinder the growth of grasses in inter tree spaces and require little care after establishment. Mixing of tree foliage with dry roughages like hay or straw improves their palatability and nutritive value. However, foliage from shrubs and trees keeps healthy animals in fair condition, but it may be inadequate as the sole feed. The chemical composition of important fodder trees are presented in Table 13 (Pal *et al.*, 1979). The scoring on the basis of most desirable traits *viz*. crude protein and on the aggregate of the scores for undesirable traits *viz*. tannin and crude fibre, considered in the reverse order has also been done. *Grewia optiva, Morus alba, Robinia pseudoacacia, Celtis australis, Artocarpus lakoocha* and *Albizia lebbek* are superior fodder trees on the basis of their chemical composition.

The shoots or sprouts, especially of tender twigs and stems of woody plants with their leaves and fruit or pods browsed to a varying extent by domestic and wild animals is referred as browse. This adds variety to their diet and helps maintain balance in the vegetation. Horse and sheep are mostly graziers and thrive on grasses and low browse; the cows, cattle and buffaloes supplement their grass diet with some browse and goats supplement its predominantly browse and pod diet with grass and herbage.

Browse play an important role as livestock feed particularly in drier tropical areas (Figure 50). Because of deep rooting habits, leathery low transpiration and xerophytic foliage, these perennials withstand extreme drought and provide foliage during scarcity. Browse can be used for day to day mixed

Table 13. Chemical composition of different fodder trees (% in dry matter).

S. No.	Botanical Name	Local name	Crude protein	Ether Extract	Crude fibre	NFE	Total ash	Insol. ash	Ca	P
1	2	3	4	5	6	7	8	9	10	11
1	*Acacia catechu*	Khair	18.72	3.05	21.88	46.69	9.66	0.76	2.46	0.14
2	*Aegle marmelos*	Bel	15.33	1.36	18.14	48.37	16.80	1.19	4.79	0.14
3	*Albizia lebbeck*	Siris	18.94	2.85	29.54	37.92	10.75	1.51	2.71	0.18
4	*Albizia stipulata*	Oee	20.99	3.35	33.74	35.99	5.93	0.60	1.03	0.14
5	*Artocarpus lakoocha*	Dheu	18.14	1.54	23.30	43.73	13.29	5.45	2.19	0.18
6	*Bambusa nutans*	Nal	14.09	1.85	32.91	35.44	15.71	8.05	0.86	0.15
7	*Bauhinia vahlii*	Taur	12.81	1.58	26.52	49.69	9.40	0.62	2.70	0.12
8	*Bauhinia variegata*	Kachnar	15.91	1.33	27.04	48.32	7.40	0.82	1.76	0.22
9	*Bombax ceiba*	Simal	12.07	4.25	23.47	50.80	9.41	0.76	3.35	0.12
10	*Cedrela toona*	Tooni	14.82	3.33	22.18	48.69	10.98	0.65	3.38	0.15
11	*Celtis australis*	Khirk	15.33	2.54	21.45	49.02	11.66	3.23	3.47	0.18
12	*Cordia dichotoma*	Lasura	12.37	1.53	26.76	41.93	17.41	3.84	3.55	0.25
13	*Dendrocalamus hamiltonii*	Maggar	18.72	1.67	29.47	36.41	13.73	7.23	0.76	0.17
14	*Eugenia jambolana*	Jamun	10.76	1.87	20.06	60.41	6.90	0.35	1.79	0.11
15	*Ficus benghalensis*	Bargad	10.29	2.76	23.12	49.55	14.28	3.00	3.45	0.12
16	*Ficus glomerata*	Gular	13.93	1.50	19.13	49.67	15.77	5.27	2.68	0.19

1	2	3	4	5	6	7	8	9	10	11
17	*Grewia optiva*	Biul	19.05	2.17	22.12	43.90	12.76	1.13	3.27	0.00
18	*Lannea coromandelica*	Kehmal	16.17	2.47	19.76	53.10	8.50	1.49	2.41	0.19
19	*Litsea glutinosa*	Rahn	14.60	2.80	27.38	47.10	8.14	1.19	2.41	0.19
20	*Morus alba*	Toot	16.64	2.30	14.38	49.70	16.98	3.32	3.62	0.23
21	*Quercus incana*	Ban	11.42	3.53	31.88	46.74	6.43	0.39	1.65	0.11
22	*Terminalia belerica*	Bahera	13.98	3.07	16.73	54.50	11.72	2.01	3.61	0.21
		Average	15.23	2.40	24.12	46.72	11.53	2.40	2.67	0.17
		Range	10.29-20.99	1.33 - 4.25	14.38 - 33.74	35.44-60.49	7.40-17.41	0.35 - 8.05	0.76-4.79	0.11-0.25
F value for df 21, 42			3.71**	2.06*	3.54*	5.58**	8.07**	7.42**	4.97**	3.92**
CD at P < 0.05			4.40	1.64	8.04	9.57	3.63	2.38	1.29	0.03

* P<0.05, **P<0.01

grazing for seasonal reserves or for drought feeding. In tropical areas, *Acacia aneura, A mellifera, Erythrina beteroana, Gliricidia maculata, Millettia thonningii* etc. are used for daily feeding. *Leucaena leucocephala* for seasonal reserves and *Acacia aneura, Grewia optiva, Celtis austrilis, Albizia lebbeck, Prosopis spicigera, Griffonia simplicifolia, Desmanthus virgatus, Cajanus cajan, Robinia pseudoacacia* are used for drought and winter feeding in different regions (Figure 51, 52 and 53).

Precautions should be taken in utilizing leucaena foliage as it contains an alkaloid known as mimosine varying from 2 to 5% which has detrimental effects on monogastric animals like horses and pigs. Sole feeding of leucaena would cause hair shedding. Thus, it is not recommended for sheep and goats. *Albizia lebbeck* contains an alkaloid at an early stage of growth. It should be fed after the leaves have matured.

The establishment of fodder trees or browse plants in natural pastures can be extremely advantageous. The tree density affects herbage yield. Relationship between herbage yield, tree foliage yield and density of mulga trees (*Acacia aneura*) at two sites in Australia have been studied by Beale (1973) and summarized in Table 14.

Table 14. Relationship between tree foliage yield (kg/ha) and density of mulga trees (*Acacia aneura*) at two sites in Southwest Queensland.

Site	Component	Tree density (Number/ha)		
		40	160	640
Boatman	Tree foliage	240	810	8330
	Herbage	2030	1150	180
Monamby	Tree foliage	180	680	2090
(Drier site)	Herbage	460	540	230

There was inverse relationship between herbage yield and number of trees per hectare. If on an average of 40-50 trees per hectare are maintained, there is little competition between the trees and forage grasses. Trees supplement natural pastures particulary during lean and stress period. Fodder trees suitable for important agro-ecological situations are listed below:

पेज 118 के बाद

Fig. 50. Hybrid Napier grown in interspaces of leucaena in a class VII land in Shivalik

Fig. 51. Biul (*Grewia optiva*) - a nutritive fodder tree for low and mid hills in Himalayas

Fig. 52. Siris (*Albizia lebbek*) - a highly adapted fast growing fodder tree for tropics.

Fig. 53. Khirk (*Celtis australis*) - a nutritive fodder tree for temperate and subtropics.

1. Hot arid and semi arid region

Khejri	(*Prosopis cineraria*)
Algaroba	(*Prosopis chilensis*)
Mesquite	(*Prosopis juliflora*)
Khor	(*Acacia tortilis*)
Mulga	(*Acacia aneura*)
Ber	(*Ziziphus nummularia*)
Safed Siris	(*Albizia procera*)

2. Hot humid sub humid

Leucaena	(*Leucaena leucocephala*)
Bamboo	(*Dendrocalamus hamiltonii*)
Khair	(*Acacia catechu*)
Mulberry	(*Morus serrata*)
Bahera	(*Terminalia belerica*)
Siris	(*Albizia lebbek*)
Gliricida	(*Gliricidia sapium*)

3. Warm humid and sub humid

Kachnar	(*Bauhinia variegata*)
Biul	(*Grewia optiva*)
Khirk	(*Celtis australia*)
Robinia	(*Robinia pseudoacacia*)
Mulberry	(*Morus serrata*)
Dhar	(*Artocarpus lakocha*)
Lasura	(*Cordia mysia*)
Siris	(*Albizia lebbek*)
Chahal	(*Anogeissus latifolia*)

4. Temperate region

Robinia	*(Robinia pseudoacacia)*
Poplar	*(Populus alba)*
Khirk	*(Celtis australis)*
Willow	*(Salix denticulata)*

5. Cold arid and semi arid

Willow	*(Salix denticulata)*
Poplar	*(Populus alba* and *P. ciliata)*
Indian poplar	*(Populus euphratica)*
Robinia	*(Robinia pseudoacacia)*

6

ARTIFICIAL GRASSLAND

Forage production from natural grassland could be improved to a great extent by replacing native grasses with a seeded mixture of high yielding and nutritive grasses and legumes. But this could be done only in plains or in areas with undulating topography where soil work could easily be done.

The anticipated life expectancy of sown pasture is generally four to five years but a good proportion are expected to remain productive for much longer. Initial successful establishment of sown pasture is important. Proper attention to the preparation of land, sowing management, fertilizer application and careful post emergence management can help to avoid establishment failure that occur.

The objective of the establishment is to achieve a lasting, satisfactory plant population which is measured on the basis of the density of tillers after establishment. A plant population of 250 to 500 plants with a about 2000 tillers/m^2 after 2-3 months of sowing is considered to be satisfactory stand.

In natural grassland, old sward should be grazed or cut as close as possible and every effort made to achieve a good seed bed. Two sprays of paraquat at a rate of 4 litres/ha of product at 7-10 days interval or a single spray of glyphosate 4-6 litres of product in 150-450 litres/ha of water are recommended for desiccating old sward before ploughing. The volume of water depends upon the density of old sward and degree of weed infestation. Competition from existing herbage must be eliminated to ensure better germination, vigour and establishment.

Development of sown pasture involves sod renovation or land tillage, fertilization, over sowing and proper management for establishment of sown species. In selecting a new grass or legume for establishing a sown pasture, their adaptation, intended use system, topography of land, availability of seed, etc should be kept in view (Figure 54 and 55).

Most of the world tropical livestock production comes from natural rather than planted pastures. Gradual transition to planted pastures and crop sources of feed may be expected in humid and subhumid zones, the semi arid and arid regions will be retained as natural grazing in varying degrees of disturbance (Humphreys, 1991).

Panicum maximum, Hyparrhenia rafa, Andropogon gayanus in South America, *Melinis minutiflora* and *Brachiaria mutica* in Latin America, *Pennisetum clandestinum* and *Digitaria decumbens* in East and South Africa, *Cenchrus ciliaris, P. maximun var. trichoglume* and *Chloris gayana* in Australia have been commonly planted in improved/sown pasture. Many reseeded grassland revert to indigenous grasses after a few years.

The successful establishment of a pasture required more skill and care than is needed for other farm crops. Techniques used are as follows (White, 1973) :

1. Destruction of previous vegetation

Destruction of previous vegetation removes competition for light, water and nutrients and gives chance for establishment of sown species. It can be done by use of herbicide, discing, tilling or burning. Desired species is sod seeded, drilled or over sown depending upon topography (Figure 56).

2. Preparation of seed bed

It is feasible only in plain or undulating topography. It can be done by conventional tillage like mould board plough or disc ploughing or by short surface working by disc harrow. In zero seed bed preparation, oversowing or drilling can be adopted. Since most of the grasses and legumes are small seeded, a fine seed bed promotes their proper germination.

3. Provision of adequate nutrients

At the time of seeding of grass-legume, starter dose of 20-35 kg P, 20 kg K_2O and 20 kg N/ha should be applied. Higher dose of nitrogen results in poor establishment of legumes. Deficiency of molybdenum and sulphur could be rectified by application of 140 g/ha of sodium molybdate and 20-30 kg/ha of sulphur through superphosphate (11% S). Drilling of 125 kg of superphosphate with seed is as effective as 375 kg/ha of its broadcast at sowing. For better establishment of legumes in particular, pH should be around 6.0 to 7.0. In case it is lower, liming should be done before sowing (Wheeler, 1950).

4. Seed of species and their mixture

The details of the adaptability of different grasses and legumes are given in Table 8 and 10. Some of the commonly used mixtures of grasses and legumes for different conditions are given below. Mixtures of different grasses and legumes have advantage of the less risk in getting uniform stand, providing higher yield, remaining green and succulent for longer period, less liable to total loss from adverse climatic condition, providing rapid and better ground cover and varied diet.

A. *Tropical semiarid*

(i) Buffel grass	*(Cenchrus ciliaris)*
Schofield stylo	*(Stylosanthes guanensis)*
(ii) Rhode grass	*(Chloris gayana)*
Alyce clover	*(Alysicarpus vaginalis)*
(iii) Marvel grass	*(Dichanthium annulatum)*
Atylosia	*(Atylosia scarabaeoides)*
(iv) Bermuda grass	*(Cynodon dactylon)*
Siratro	*(Macroptilium atropurpureum)*
(v) Klein grass	*(Pennisetum coloratum)*
or	
Blue panic	*(Panicum antidotale)*
Townsville stylo	*(Stylosanthes humilis)*

(vi) Daryai grass	(*Andropogon pumilus*)
Atylosia	(*Atylosia scarabaeoides)* or
Shrubby stylo	(*Stylosanthes scabra)*

B. *Tropical humid*

(i) Guinea grass	(*Panicum maximum)*
Siratro	*(Macroptilium atropurpureum)*
(ii) Setaria	*(Setaria anceps)*
Green leaf Desmodium	*(Desmodium intortum)*
Lablab bean	*(Dolichos lablab)*
(iii) Para grass	*(Brachiaria mutica)*
Silverleaf Desmodium	*(Desmodium intortum)*
(iv) Broadleaf paspalum	(*Paspalum wettsteinii)*
Centro Calapo	*(Centrosema pubescens* or *(Calapogonium mucunoides*)
(v) Carpet grass	(*Axonopus affinis*)
Glycine	(*Glycine javanica)*
(vi) Molasses grass	*(Melinis minutiflora)*
Axillaris	*(Macrotyloma axillare)*
(vii)Rhode grass	*(Chloris gayana)*
Puero	*(Pueraria phaseoloides)*

C. *Subtemperate*

(i) Kikuyu grass	(*Pennisetum clandestinum*)
White clover	(*Trifolium repens)*
(ii) Dallis grass	(*Paspalum dilatatum or*
or Plicatulum Silver leaf desmodium	*Paspalum plicatulum*) (*Desmodium uncinatum*)
(iii) Tall fescue	(*Festuca arundinacea)*
Birdsfoot trefoil	(*Lotus corniculatus*)

Kenya white clover	(*Trifolium semipilosum*)
(iv) Setaria	(*Setaria anceps*)
Subterranian Clover	(*Trifolium subterraneum*)
(v) Lunji	(*Themeda anathera*)
Lespedeza	(*Lespedeza sericea*)

D. *Temperate moist*

(i) Perennial rye grass	(*Lolium perenne*)
White clover	(*Trifolium repens*)
(ii) Timothy	(*Phleum pratense*)
Red clover	(*Trifolium pratense*)
(iii) Orchard grass	(*Dactylis glomereta*)
White clover or	(*Trifolium repens* or
Lucerne	*Medicago sativa*)
(iv) Tall Fescue	(*Festuca arundinacea*)
Sainfoin	(*Onobrychis viciifolia*)

In high rainfall areas, 2-3 kg/ha birdsfoot trefoil (*Lotus corniculatus*) and on saline soil, strawberry clover (*Trifolium fragiferum*) at the rate of 3 kg/ha should be added.

E. *Temperate dry*

1. Crested wheat grass	(*Agropyron cristatus*)
Lucerne	(*Medicago sativa*)
2. Smooth brome grass	(*Bromus inermis*)
White or yellow sweet clover	(*Melilotus alba*) *or* (*M. officinalis*)
3. Red fescue	(*Festuca rubra*)
White clover	(*Trifolium repens*)

White clover is by far the most important species to be introduced under most situations in temperate region and should be included in mixture. In low rainfall areas, subterranian clover may be most favourable. Red clover is useful in drier areas for summer and autumn production; however, its persistence is low. Introduction of lucerne by sod sowing and over sowing is most desirable species in semi arid temperate region. In steep and rocky terrain where over drilling is not possible, over sowing can be done successfully (Semple, 1970; Langer, 1973).

5. Time and method of sowing

The presence of adequate moisture is essential for sown seed to germinate. Thus in tropical and subtropical conditions, ideal time of seeding is the onset of monsoon. Early monsoon is suitable time for planting of sprigs, rooted slips, runners and cutting of perennial grasses under tropical condition. In temperate region, seeding of grasses and legumes can be done either in early spring on melting of snow, or in early autumn when there is usually adequate moisture in soil. Temperate forages can be sown either in spring on melting of snow or in early autumn if sufficient moisture is there in soil (Singh, 1986).

Shallow seeding of grasses and legumes is generally done because of their small seed size. Seed depth should vary 1.0 to 1.5 cm for small seeded and 1.5 to 2.5 cm for larger seeded grasses and legumes. Broadcasting and drilling is done for sowing of pasture. Forage species differ markedly in their seed weight and germination. A figure for population commonly used is 10 million seed/ha. An even establishment of 50 per cent of seed sown should be adequate for satisfactory stand. Seed rate of different grasses and legumes to be used are given in Table 9 and 11. Twenty to 30 kg seed of grass and legume per hectare is considered adequate. Grass seeds could be broadcasted, provided they are harrowed in after sowing to ensure an adequate cover. The direct over seeding into natural or native grasslands usually results in poor stand and reversion to original type in few years because of competition (Sucking, 1965).

पेज 126 के बाद

Fig. 54. Setaria-desmodium grassland in mid hills of H.P.

Fig. 55. Phalaris established in Kullu Valley

Fig. 56. Holstein-Friessien cows grazing on an artificial grassland in U.K.

Fig. 57. Artificially established ryegrass-white clover grassland in U.K.

Grass seed coating with various combinations of lime, reverted superphosphate and nitrogen has favourable effect on their establishment on oversowing in native pasture. Depending upon the situation, techniques such as burning, grazing, trampling, or herbicide treatment may be used to reduce the competition of existing vegetation with sown species. Paraquat is commonly used herbicide to reduce competition by indigenous flora. It has little residual effect (Whiteman, 1980).

6. Early management

Early proper grazing management affects the establishment of all sown species. Frequent light grazing in first growth season helps in establishment of slow growing species (Figure 57). Sheep or young calves are better livestock than cattle for early grazing.

Weeds are often a problem in sown pastures. Various herbicides such as MCPB can be used for their control but in most cases the "hoof and tooth" treatment using high stock numbers for a short period is still the most effective and is much cheaper. The harvesting of hay, silage or a seed crop in first year of pasture is detrimental to legume establishment.

7. Correction of nutrient deficiencies

Most soils in grasslands are deficient in nitrogen which can be corrected through vigorous legumes introduction which in turn is affected or limited due to sulphur, phosphorus and molybdenum deficiency. If soil is acidic, lime pelleting of seed is often sufficient to improve the soil environment for lucerne inoculation and give seed nodulation without further liming. Heavy broadcast of liming is uneconomical in steep hilly terrain (Sampson, 1952; Serviss and Ahlegren, 1955; Donahue *et al.*, 1956).

7

FORAGE CONSERVATION

Herbage growth in grassland is generally for 3 to 4 months only. During this period there is surplus of green forage, and there is scarcity during winter and dry summer months. To ameliorate the problem of green forage availability during lean period, attention need to be given for the conservation of surplus fodder during monsoon season. Conservation of surplus forage helps in grassland management also by adjustment of the area available for grazing. Surplus forage can be conserved in the form of hay and silage with minimum loss of quality and quantity (Watson and Nash, 1960).

A. HAY

Forage crops cut before it is ripe and thoroughly dried is called hay. Hodgson *et al.* (1948) defined hay as a forage dried so as to retain most of the leaves, without deterioration of dry matter and nutrients, without mould development, having its natural green colour and palatability, and capable of being stored over a long period of time (Figure 58). The good quality hay is characterized by leafiness, bright green colour, pleasant aroma, pliable texture, high nutritive value, palatability and freedom from foreign material (Cherry, 1968). The quality of hay depends upon the species that is the type of grass and legume, stage of harvesting and the weather likely to prevail during drying, curing process and storage (Figure 59).

Forage species have inherent characteristics of their chemical composition. Legumes are rich in protein and grasses in carbohydrates. In general all herbage species are rich in nutrients at early stage of growth and it declines with the advance in growth. However, dry matter production is low at early stage and high at full growth stage. In order to get good quality hay, a compromise has to be made so that there is high

Fig. 58. Grass bundles for hay making

Fig. 59. Stack of local grass hay and paddy straw

dry matter yield with good quality. Generally grasses should be cut at full flower stage. The weather at harvesting and during drying plays an important role in preparation of hay. Bright sunny weather results in quick drying of mowed grass without much loss of dry matter and quality. If it gets wet during drying, there is loss of nutrients and dry matter, growth of moulds and there is spoilage. For proper drying of mowed grass in field, it should be turned over, heaped up and shaken several times until drying is complete. The moisture content in sun dried good hay should not exceed 15-20 per cent.

Nutrient losses may be 20 to 30 per cent in hay making compared to 10 to 15 per cent in silage. Hay is more nutritious and palatable than straw because it is cut and dried before maturity. A grass or legume that is fit to be fed to livestock can be converted into hay by proper drying, certain species are better suited than others. Forage plants that are best suited for hay making are thin stemmed and leafy and remain sufficiently nutritious and palatable after drying (Narayanan and Dabaghao, 1972).

Forages must be cut at an early vegetative stage to get good quality hay. When cut early, the ratio of leaf to stem is high. Leaves have a greater nutritive value and a lower fibre content than stems. Early cut forage will therefore be high in protein and total digestible nutrients and low in fibre. Mature plants give a greater tonnage but the leaf to stem ratio is much lower in nutritive value and palatability. The right stage for hay is a compromise between the best quality and highest yield. In general, good yields of reasonable quality hay are obtained when legumes are cut at the bud to early bloom stage and grasses at or immediately after heading out.

Curing: Newly mown forage must be reduced from a moisture content of 70 per cent or more to 25 per cent or less before it can be safely stored. Drying must be rapid to avoid excessive losses of dry matter, vitamins and soluble nutrients. Leaves of grasses and legumes contain as much as 70 per cent of protein and 90 per cent of carotene found in entire plant. The greatest loss of nutrients occurs when leaf loss is high.

Drying can be done in grassland, ground, road side or roof top. Apart from the obvious loss of moisture, many chemical changes also occur. Plants continue to respire even after cutting till leaves get wilted. Thus there is loss of dry matter and carbondioxide. Protein content also decreases because of enzyme activity in tissues. Hay should be stored in barns or shelter preferably after baleing (Figure 60) to prevent loss due to weathering.

Fig. 60. Baleing of hay (Kaul & Gogte, 1993)

Following generalization has been made about hay made from tropical and subtropical forages (Crowder and Chheda, 1982).

1. Although hay made from natural grassland is rarely of high quality, it has a better feeding value than the standing hay or foggage left in the field.
2. Hay made from the first growth after rains or from the aftermath is considerably superior to hay made from the whole seasons growth or hay made during the dry season.
3. Crude protein and digestible crude protein contents of tropical grass hay range between 4-10 and 1-6 per cent, respectively.
4. Crude protein and digestible crude protein contents of legume hay are high and range between 11-17 and 7-13 per cent, respectively.

Before storage whole plant hay should be reduced to a moisture content of 25 per cent, and chopped or baled hay to 20 per cent. At moisture contents above these levels, biological activity continues resulting in destruction of carbohydrates and vitamins, decreased digestibilty of proteins and in extreme cases, complete destruction occurs due to spontaneous combustion if stored in barn.

Hay is made mostly after monsoon rains are over in September and October and sun is sufficiently bright to dry the plants in the swathe in 2-4 days. At very high altitude in temperate region, grasses from meadows are harvested in August, dried and stacked on trees or over roof top for feeding during winter.

When hay is cured in field, a loss of about 15 to 20 per cent in dry matter is inevitable, and under adverse condition it may increase up to 90 per cent or even 100 per cent. Average losses of dry matter during hay making and storage is around 25 to 30 per cent. The loss of 36 to 60 per cent in starch equivalent, 25 to 60 per cent in digestible protein and nearly whole of carotene takes place during curing. If the weather is wet leaching, bleaching, leaf shedding, mould growth may account for losses up to 40 to 60 per cent. If the moisture content is high at stacking time, fermentation may start causing considerable loss of nutrients and dry matter.

Different grades as given below in Table 15 are used to indicate the quality of hay in United States (Lancaster *et al.*, 1949).

Table 15. Grade of hay and their quality

Grade No.	Leafiness, Alfalfa	Minimum per cent of leaves Lespedeza	Colour
1	40	45	Green to greenish brown
2	25	30	Greenish brown to brown
3	10	10	May be brown

Sample grade Hay which does not come within the requirements of any of the numerical grades or which contains more than a trace of injurious foreign material; or which has any objectionable odor; or which is under cured, heating, hot, wet, musty, moldy, caked, badly broken, badly weathered, badly frosted, badly over ripe, or very dusty; or which is otherwise of distinctly low quality.

The words Extra Leafy, Leafy, Extra Green, Green or Coarse are added to Grade No. to denote other characteristics of hay.

Composition of hay

The composition of ordinary hay as compared to the green grass is given in Table 16 (Aiyer, 1962).

Table 16. Composition of hay and grass

	Water	Albuminoids	Fats	Carbohydrates	Fiber	Ash
Hay	15.0	9.4	2.6	38.8	28.5	5.7
Grass	68.8	3.4	0.9	14.0	10.8	2.0

The composition of hay varies with species also (Table 17).

Table 17. Chemical composition of certain grasses

Species	Crude protein, %	Crude fiber, %	N. free extract, %
Cynodon dactylon	6.8	24.2	56.0
Dichanthium annulatum	5.2	38.5	44.5
Chloris gayana	12.7	27.3	45.8
Heteropogon contortus	5.6	38.2	46.0

Measurement of hay quantity

The quantity of hay is determined by measuring the volume of hay stack and then dividing the volume by cubic feet value required per tonne of stack hay. On an average, value of 512 cubic feet per ton may be used to estimate the weight of stored loose hay. For baled or chopped hay, this value is about 250.

To determine the volume of various shaped stacks, four measurements (in foot) are made *viz.*, length (L), width (W), over (O) and circumference (C). Over is the distance from the ground on one side over the stack to the ground on the other side. Width is taken at ground level. If there is variations in the length of a stack, two or more measurements are averaged. The following formulae are used for measuring the volume of stacks:

(a) Square, flat topped stacks :

(0.56 x O) - (0.55 x W) x (W x L)

(b) High, round topped stacks :

(0.52 x O) - (0.46 x W) x (W x L)

(c) Low, round topped stacks :

(0.52 x O) - (0.44 x W) x (W x L)

(d) Round stacks :

(0.04 x O) - (0.012 x C) C^2

B. SILAGE

Silage is a moist succulent feed produced as a result of controlled fermentation of fresh forage when stored in a silo under anaerobic conditions. The process of making silage is called ensiling. Any herbage or forage crops can be made into silage provided it has desirable level of dry matter content and soluble carbohydrate.

When green plant material, whole or chaffed to the length of 2 to 2.5 cm is stored where it can ferment, it develops acid which acts as preservative for the nutrients of the forages ensiled.

Silage can be prepared in pits, trenches or plastic bags (Figure 61, 62 and 63). Pits are dug 2 to 3 m deep and 1.8 to 2.4 m wide at high level locations where the water table is more than 3.7 m below the surface. Trenches can be of varying width and length. Uncemented floor and sides of pits and trench should be covered with 10-15 cm layer of dry grass or any other suitable material.

The principle in making silage is to keep the green fodder material tightly packed in impervious containers excluding air as much as possible. Grass should be harvested at the right stages of growth *viz.* between the flowering and milk stage. The important conditions for getting good quality silage are :

(a) Storing the plant material at moisture content of 65 to 75%. If the herbage is too succulent, it is slightly wilted to reduce the moisture content. There are three possible ways of

Fig. 61. Silage in plastic bags

Fig. 62. Trench silo dug out on a hillock slope

Fig. 63. Trench silo filled with chaffed forage and sealed with plastic sheet. Discarded vehicle tyres placed on top for compaction. Plastic silos are in fore ground

finding out moisture content of herbage (Lancaster *et al.*, 1949). These are: (i) The Grab Test (ii) Oven Test, and (iii) Scientific Equipment Test. The grab test is carried out by merely grabbing a handful of the chaffed forage and squeezing it in the fist. If water drips from hand after squeezing, the material is too wet. If the material holds its shape after squeezing, it is still too wet. If on opening the fist after squeezing, the ball of the material is compact, but not wet and shiny, and then slowly springs apart, the moisture content in herbage is just right i.e. between 65 to 70%. If the ball does not form and the material flies apart when released, after squeezing, it is too dry.

In oven test, one kilogram chaffed sample of forage may be dried for 6 to 8 hours in the oven of a kitchen or laboratory at 100°C. There should be vent to permit moisture to escape. After drying, whole sample should be weighed again to determine dry matter percentage. Moisture meters of different types are available which could be used to determine moisture content in herbage, but are expensive for use of farmer.

(b) Excluding air when stored. For this herbage is chaffed and packed and trampled.

(c) Encouraging a rise of temperature from 30 to 38°C. For this proper compaction and sealing is done in silos.

One cubic metre of space in silos is provided for 400 kg of fodder, or a cubic foot will hold about 11 kg of silage. The density of silage is influenced by silage depth, diameter of silo, per cent moisture and crop ensiled. Pits and trenches are made wider at the top than at the bottom as this helps the silage to pack tighly against this sides as it settles. The capacities of different types of silos are given in Table 18 (Lancaster *et al.*, 1949).

Table 18. Different types and sizes of silos and their capacities

A. Trench silos			
Depth, ft.	Bottom width, ft.	Top width, ft.	Weight per Lineal ft., lb.
4	5	7	840
4	6	8.75	1015
4	7	10.25	1155
B. Tower silos			
Depth of silage, ft.		Inside diameter, ft.	Capacity, tons
20		10	27
24		10	34
24		12	49

The chaffed grass material is filled in trench or pit uniformly and trampled well at each layer. Filling is done about a meter above ground level and top is rounded and sealed up with polythene sheet and covered with clayey soil. If there is aeration, it promotes the activity of moulds which spoils the silage and make it unpalatable. The respiration in packed herbage continues for a short time. As the temperature goes up, the bacteria present become active and cause fermentation using carbohydrates. Consequent to this carbon dioxide, organic acids like lactic, acetic and butyric acids and small quantities of alcohol are produced.

When herbage ensiled contains 65 to 75 per cent moisture and sufficient sugar, anaerobic lactic acid bacteria become active and produce a good, clean smelling silage of high quality. However, if the herbage ensiled is too rich in proteinaceous substances, the butyric acid type fermentation will predominate. Since butyric acid has a sharp, disagreeable odour, the silage produced under this kind of fermentation is not relished by livestock.

If the herbage to be ensiled is too succulent or does not contain adequate soluble carbohydrate, addition of preservatives is required for making good quality silage. If sugar content is less than 2 per cent of fresh weight, a high rate of additive will be needed to ensure a good fermentation. Legumes have very low sugar content. The preservative add extra carbohydrate, lower the pH value and inhibit the action of butyric acid type of bacteria.

The preservatives may be material with high sugar content such as molasses, cereal grains, beet pulp etc. or chemical additives such as sodium metabisulphite or sulphurdioxide, modify the fermentation process and reduce the smell.

Silage preparation is satisfactory if herbage has high dry matter and adequate sugar content. Soluble carbohydrate content in temperate grasses ranges between 13 to 16 per cent of dry matter whereas in tropical grasses, is often below 6 per cent. When material is low in dry matter and sugar content, additives are required to be used to ensure satisfactory preservation (Table 19). A wide variety of additives are used and their ingredients fall into three main classes: fermentation stimulants, fermentation inhibitors and inhibitors of post opening deterioration.

Carbohydrate sources such as molasses and bacterial cultures (inoculants) which supplement the naturally occurring population of lactic acid bacteria act as fermentation stimulants. Fermentation inhibitors may be of two kinds. First, those which bring about qualitative changes in the microflora. These are mainly acids which rapidly lower the pH so that clostridial activity is restricted and the lactic acid bacteria stimulated. The most effective material in this group is formic acid. The second type are those which inhibit the microflora in general. Their use results in sterilization of the crop. Sodium nitrite and formaldehyde are commonly used microflora inhibitor. Propionic acid is currently pre eminent inhibitor of post opening aerobic deterioration of silage but appears to be uneconomic at levels of usage.

Table 19. Silage additives

Name	Ingredient	Quantity tonne herbage
Add F	85% Formic acid	2 -5 litres
Advance	Lactobacillus plantarum & other micro-organisms enzyme	0.5 -1 kg.
Balegard	Propionic acid, Sodium benzoate, mould inhibitors surfactants	1 - 2 litres
Cosil	Sulphuric acid	2 - 3 litres
Farmline silage additives	25% Formic acid 35% Sulphuric acid 20% Formalin	3 - 5 litres
Fodder-guard 'S'	Formic acid, Propionic acid, antioxidants, mould inhibitors	1.5 - 2.0 litres
G A P	Phosphoric acid, propionic acid, alcohols, antioxidants	1 litre
Hy-Acid 2000	50% Formic acid 20% Sulphuric acid 1% Phosphoric acid	2.5 - 4.0 litres
Molasile	75% Molasses 25% Cider apple vinegar	4.5 - 9.0 litres
Molasses	50% Sugar	9 - 15 litres
Progard	20% Formic acid 17% Sulphuric acid 40% Formalin	2-3 litres
Scotsil	Sulphuric acid	2.3 litres
Silagard	50% Formic acid 20% Sulphuric acid 1% Phosphoric acid	2.5 litrs
Silagepak 'P'	Micro-organisms	1.0 kg
Silage F 100	36% Formic acid 64% Formalin	4.0 litres
Silaphos	20% phosphoric acid 40% Sulphuric acid	1.0-1.5 litres
Silasafe	Ammonium tetraformate	2 -5 litres
Silo-Action	Lactobacilli enzymes, antioxidants, mould, inhibitors	0.11 kg

Sure Sile	Lactobacilli, Streptococci enzymes	0.5 kg
Sylade 2	20% Sulphuric acid 50% Formalin	2.5 - 5.0 litres
Sylextra	17% Sodium nitritie 6% Calcium formate	2.3 litres

* Source : Donaldson and Edwards, 1983

Dilute sulphuric and hydrochloric acids or phosphoric acid are also sprayed on chaffed herbage at the time of ensiling to reduce pH value from original 6 to 3-4. This suppresses bacteria activity and herbage gets converted into silage without too much damage. Addition of preservative improves the ensiling process.

The preservative can be added according to herbage material (Table 20)

Table 20. Type of herbage material and the quantity of preservative

Herbage material	Preservative per 100 kg of herbage
Grass alone	2.5 - 3.5 kg of molasses or 4.5 to 6.8 kg of kibbled grain or 4.5 kg of beet pulp
Legume and grasses	3.5 kg of molasses or 6.8 to 9.0 kg of kibbled grain or 4.5 or beet pulp
Legume alone	3.5 to 4.5 kg of molasses or 6.8 to 11.3 kg of kibbled grain or 6.8 kg of beet pulp

Salt can also be added at the rate of 0.1 per cent of herbage material. Addition of urea 0.02 per cent in pure grass herbage improves the quality of its silage. Sodium metabisulphite, the most effective chemical preservative, is added at a rate of 0.4 kg per 100 kg of herbage material.

The fermentation in silo takes place during first two or three months, and after this silage remains practically unchanged for another 12 to 18 months. Silos can be opened after three months of sealing. Once opened, silage should be consumed.

Loss of nutrients

In the process of ensiling, there is loss of dry matter and the nutrients. This may vary from 10 to 40 per cent depending upon the silo, herbage ensiled, preservative used and the method of ensiling. Under average condition, the loss is around 10 to 20 per cent. In earthen silo pits or trench, loss of 25 to 35 per cent may occur due to spoilage and rotting.

The quality of silage depends on the stage of maturity, dry matter and nutritive content, particulary the carbohydrate fraction and the process of ensiling. Good silage has a clean odour uniform in greenish colour without any mould sliminess or mushy rot. The dark brown colour indicates excessive heating and if the colour is black, the silage is considered rotten and useless. If the herbage is containing less than 20 per cent dry matter, it gets composted easily and butyric acid fermentation takes place which makes the silage unacceptable

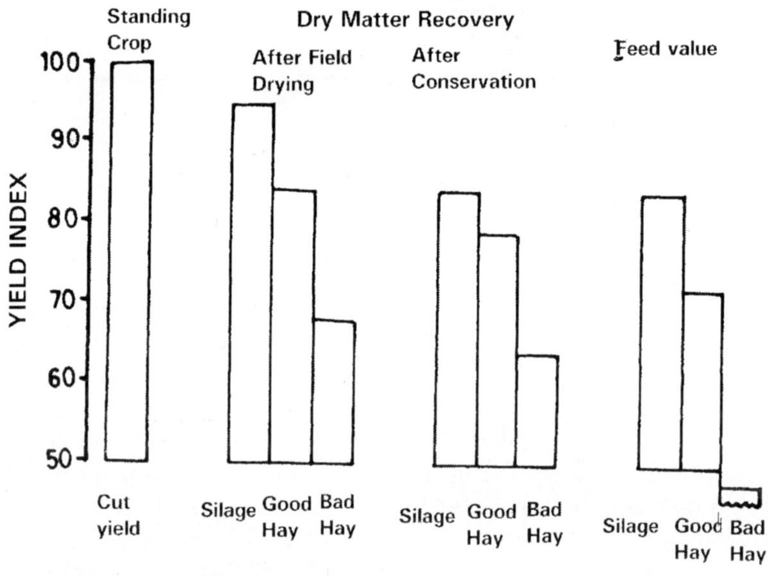

Fig. 64. Effect of conservation system on recovery of dry matter and relative feeding value

to livestock. Over ripe herbage material is difficult to compact and leads to developments of moulds and high temperature which produces musty silage unacceptable to livestock.

Conservation system affects recovery of dry matter and feeding value (Figure 64). There is higher reduction in dry matter as well as feeding value in hay preparation compared to silage. Field curing of the hay reduces feed value especially in rain condition (Bad hay). Hay stubbles are slow to regrow. Thus, hay making not only fails to produce good quality winter feed but also reduces the grass yield for the whole season.

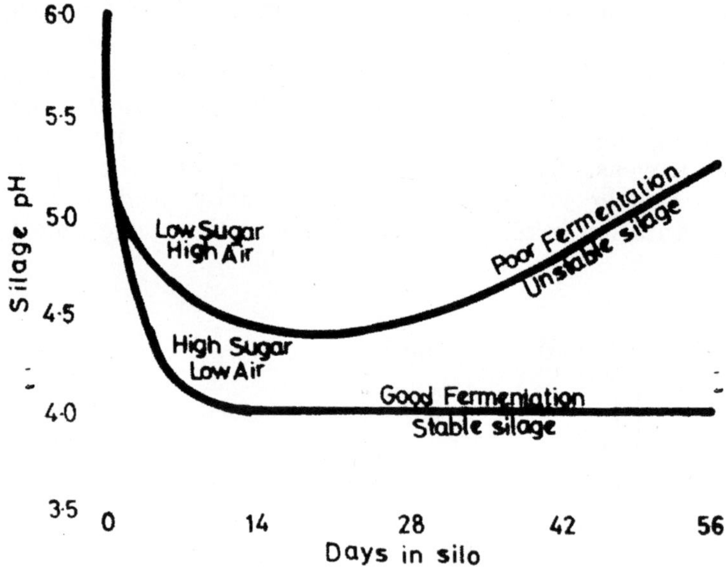

Fig. 65. Pattern of silage pH in well and poorly preserved silages

Good silage is produced on rapid conversion of sugar to lactic acid. Air in silo or lack of sugar in grass results in wrong type of fermentation (Figure 65). Bacteria develop which break down lactic acid to weaker butyric acid and protein is converted into less valuable non-protein nitrogen. Silage pH rises and losses increase .

8

HERBAGE QUALITY

Grass together with legumes and other broad leaved plants is collectively referred to as herbage. Its quality depends upon its chemical composition, the digestibility of its constituents, how much of it is eaten and absence of antiquality components. In general, grass is natural feed of animal, but it is not a balanced feed for animal production purpose. Some herbage may have high energy content but may contain insufficeint protein or vice versa. The productivity of animal not only depends on herbage quality but also on animal potential carrying capacity, grazing pressure and provision of other supplement (Crowder and Chheda, 1985). The herbage quality in true sense can be determined only by means of feeding trials. However, broad idea can be made on the basis of their chemical composition (Lander, 1949; Mcllroy, 1972).

Chemical Composition

The basic chemical composition of plant or herbage is shown in Figure 66. They contain a large amount of water varying from 90 per cent in young leafy stage to about 15 per cent at maturity. Thus, the proportion of dry matter in fodder may vary from 10 to 85 per cent depending upon growth stage, species and environmental conditions. The dry matter is composed of organic matter and ash content (Crampton and Harris, 1969). The proximate analytical scheme resolves a given feed stuff into five fractions : crude protein (CP), fat or ether extract (EE), crude fibre (CF), ash and nitrogen free extract (NFE).

The sum of the crude fibre and nitrogen free extract represents the total carbohydrate of fodder. The ether extract represents the fat or oil fraction including fat soluble vitamins.

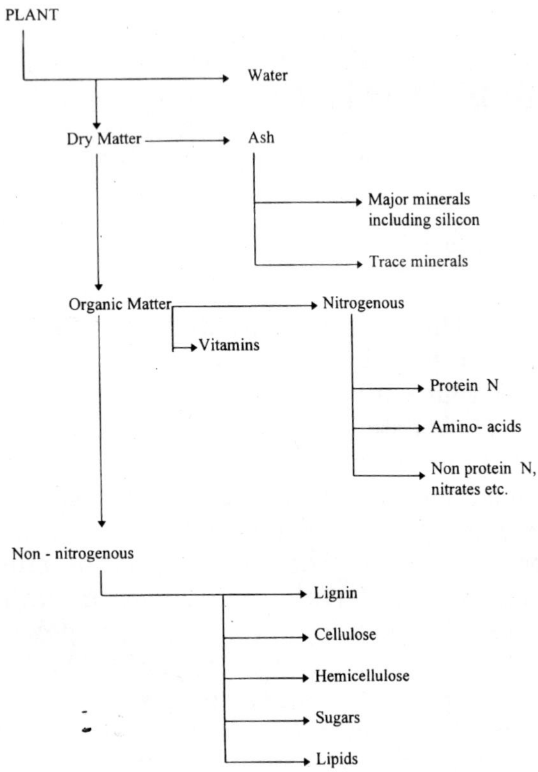

Fig. 66. Chemical composition of a plant

It is usually in the range of 1-3 per cent and somtimes upto 5-6 per cent. Crude fibre is the proportion of the total carbohydrate of the plant and includes cellulose and lignin. It is insoluble in dilute acid and alkali and varies from 20-25 per cent in young plant to 30-40 per cent in mature plants. Ninety five per cent of the crude fibre fraction is cellulose which is converted mostly to acetic acid by microorganisms in the rumen during the process of digestion. The nitrogen free extract is determined as the difference between the original weight of a dry matter sample and the sum of the weights of ether extract, crude protein, crude fibre and ash and it includes sugar, starch, some hemicellulose and most of the lignin. It is an index of the non-cellulose portion of carbohydrate and makes up about 35 to 55 per cent of the dry matter of fodder. Crude protein is determined by multiplying the nitrogen fraction by a factor of 6.25. It ranges from 3 to 20 per cent or higher in young plants. It includes all nitrogenous substances, mineral N, amide N, amino acids and proteins. Only about 60% of the potential energy of protein becomes available to the animal and the remainder is lost as urea in urine and as undigested protein. Ash content is highly variable not only in its total amount but in its component parts and is a poor index of any of the inorganic nutrients. Some fodder may be high in silica which is of no nutritional value (Gohl, 1975).

An alternate method of partitioning of chemical composition was proposed by Van Soest (1966). According to this, herbage is made up of two basic dietery fractions : cell wall content (CWC) and cell content (CC).

Cell wall content

Cellulose, hemicellulose, lignin and silica which form about 30-80 % of plant dry matter are the principal cell wall components. The proportion of these constituents increases in plants with the advance in growth. In general, grasses contain a higher percentage of cell wall content in comparison to legumes. The quality of fodder depends upon the content and digestibility of these constituents. Out of these, cellulose is more digestible than hemicellulose whilst lignin is virtually indigestible.

Cell content

Carbohydrates, fats, proteins, organic acids, minerals and vitamins constitute the cell contents. Carbohydrates are the most common constituent of plant dry matter. They include compounds varying from simple sugars like glucose, fructose and sucrose to polysaccharides like starch, fructose, cellulose and hemicellulose. Sugar refers to water soluble carbohydrates, which consist of glucose, fructose, sucrose and fructosans. Grasses in general contain a higher sugar content than legumes. It may vary from 4-12 % in legumes and 8-24 % in grasses depending upon species and growth stage. Starch which is composed of a large number of glucose molecules is mostly found in legumes varying from 5-8 % of dry matter. Grass foliage contains only trace of starch (Chopra *et al.*, 1956; Dogra *et al.*, 1979).

Lipids form less than 3 % of dry matter and are composed of fats or oils and fat soluble vitamins A, D and E or their precursors. Protein constitute about 80% of the total nitrogenous compounds of plant dry matter, the remainder being ammonium salts, nitrate and amides. The crude protein content varies from 2.4% in wheat straw or mature grass to 25% in young grasses and legumes. Leaves in general contain more protein than stems. Seven to eight per cent crude protein is considered to be a requirement for normal microbial synthesis of rumen organisms.

Elements other than carbon, hydrogen, oxygen and nitrogen form the mineral content of plant. It varies from 3-13% of dry matter. Legumes are richer in calcium, magnesium, iron, copper, cobalt, molybdenum and manganese than grasses. Emission spectrometers (quantometers) are available which read out simultaneously quantity of 24 elements in herbage samples.

Digestibility

The value of fodder depends not only on its chemical composition but also on its digestibility as it is this digested material which becomes available to the animal for assimilation and utilization. The proportion of feed or fodder which is digested as it passes through an animal body is referred to as the digestibility. Digestibility indicates the energy value of a feed and is expressed in terms of digestible organic matter which is referred as 'D' value.

The digestibility of herbage organic matter is calculated on the basis of feeding trials as follows:

Digestibility of organic matter % =

$$\frac{\text{Weight of O.M. eaten} - \text{Wt. of O.M. in faeces}}{\text{Wt. of O.M. eaten}} \times 100$$

Herbage under test is fed to experimental animals for a preliminary period of 7 days followed by feeding during trial period of 10-12 days during which observation is recorded.

Reliable estimates can also be made through in vitro digestibility analysis such as that of the two stage in vitro technique (Tilley and Terry, 1963). Apparent digestibility of dry matter can be estimated on the basis of chemical composition also (Raymond, 1969). Digested organic matter and ash content form the total digestible dry matter.

At an early growth stage, digestibility is high and it starts decreasing with the advancing growth. After flowering digestibility falls rapidly. The D-value at early and leafy growth stages in most grasses may be 70%, whereas, at maturity it drops to 60% or lower. The D-value of a plant depends upon the species, genotype and growth stage. In the immature forages, the stem is more digestible than other components. The digestibility of the leaf fraction, however, decreases slowly in comparison to stem with the advancing maturity. Within a particular forage species at similar stage of morphological development, late varieties tend to be less digestible than early varieties. Applied nitrogen increases crude protein but decreases soluble carbohydrate content. It thus has no effect on the digestibility of herbage. Of all the chemical constituents, sugar, proteins and fats are digested rapidly. Cellulose and hemicellulose are degraded slowly, whereas, lignin is indigestible. The digestibility of grasses, crop residues and by products varies from almost zero to 80 per cent (Figure 67), largely because of variation in their cell wall content (McDowell, 1985).

The chemical compounds within cells can be grouped as shown in Figure 68 according to their alcohol solubility and digestibility by ruminants (Pearson and Ison, 1987).

The cellulose in its pure form is 100 per cent digestible and is less accessible to degradation when it is bound within plant tissues. Hemicellulose which is bound into the cell wall and protected by coatings of lignin to various extents, has a variable but lower digestibility. Cell wall material in young herbage has a digestibility of as much as 90 per cent but this falls to 30 per cent as the cell becomes increasingly lignified with age.

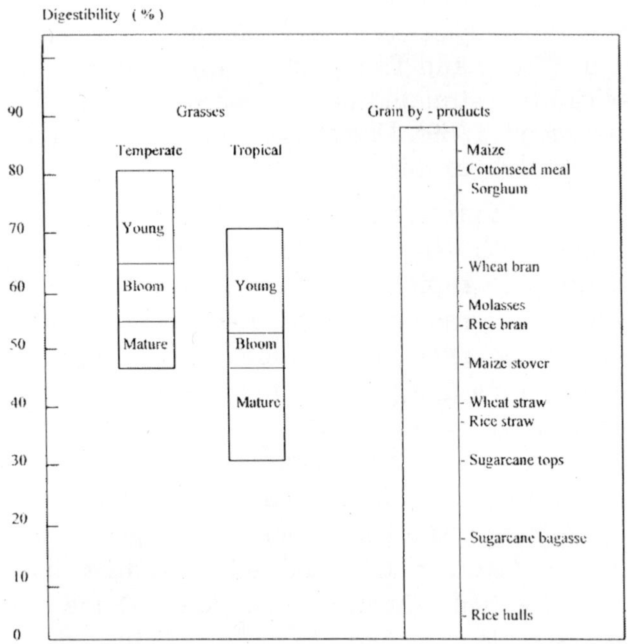

Fig. 67. Digestibility of grasses and grain and byproducts

Total Digestible Nutrients (TDN)

It is the sum of the digestible protein, carbohydrate and the digestible fats multiplied by 2.25. The value of TDN and DCP are commonly used for computation of rations for farm animals (Sen and Ray, 1964; Mays, 1974). In herbage, digestible crude protein varies from nil in wheat and rice straw to 15 per cent in lucerne hay (fresh basis) and total digestible nutrients from 14 to 15 per cent in most of green fodder to 60 per cent in lucerne hay. DCP and TDN in concentrate mixture varies from 16 to 20 per cent and 70 to 75 per cent, respectively, on dry matter basis.

Utilization of energy

Animals do not use all the energy in the feed that is eaten. A schematic representation of the recognized partition of energy is given in Figure 69.

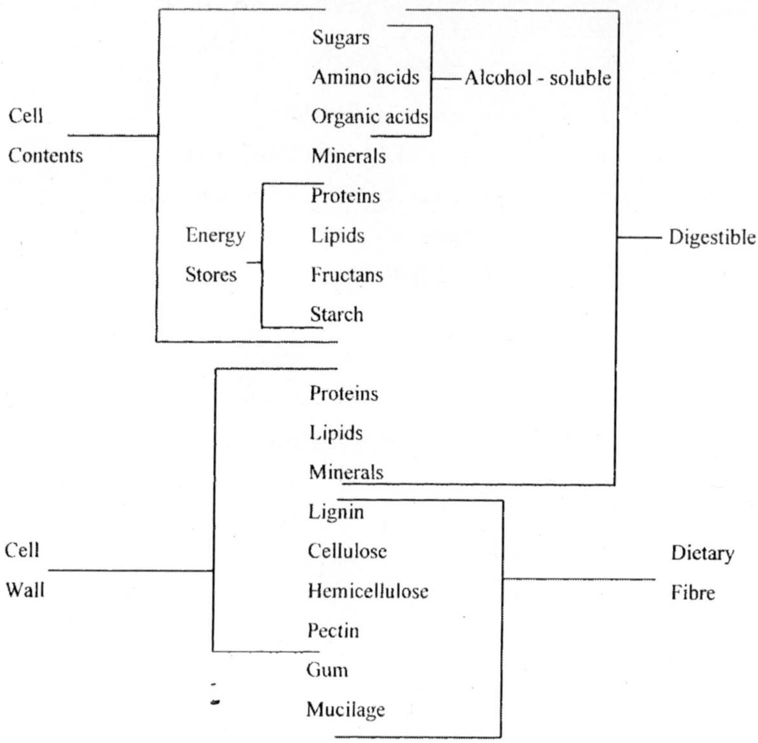

Fig. 68. Chemical compound with in cells according to alcohol solubility and digestibility

The following desirable minimum quality requirements from grassland for maintenance and production has been suggested (Leng, 1987).

(i) The digestibility of feed should be high, i.e. in excess of 65 per cent.

(ii) Soluble protein in the diet should provide about 30 g N per kg digestible dry matter.

(iii) 'Escape' dietary protein should provide about the same protein as arises from rumen microorganisms, i.e. about 200 g protein per kg digestible dry matter intake.

(iv) The level of fat (lipids) in the diet should approach that of temperate grasslands, i.e. 4-8 per cent of dry matter.

GROSS ENERGY OF FEED (GE)

→ Faecal energy

  Feed origin

  Metabolic (animal) origin

APPARENTLY DIGESTIBLE ENERGY (DE)

→ Gaseous products of digestion, principally methane

→ Urinary energy

METABOLIZABLE ENERGY (ME)

→ Heat increment

  Heat of fermentation

  Heat of nutrient metabolism

  Temperature regulation

NET ENERGY (NE)

→ Production energy (NEp) for growth , fattening, milk, reproduction, wool, work

Maintenance energy (NEm)

- Basal metabolism
- Activity

Fig. 69. The fate of the energy with in feed

(v) Glucogenic energy should be high proportion of the volatile fatty acid energy absorbed from the rumen, i.e. greater than 25 per cent.

Antiquality constraints and disorders

Herbage though rich in nutrients may cause disorder in animal due to presence of certain alkaloid or deficiency or excess of certain chemical constituent. Important disorders are as follows:

Bloat:

It is observed in ruminant animals grazing succulent legumes and is due to formation of a stable foam in rumen. It is commonly observed in temperate countries. Tropical pasture legumes in general do not normally contain the bloat inducing 18 S protein. Gases normally lost by belching are retained and pressure is increased in the left chamber of the rumen resulting in inhibition of eructation mechanism. When this occurs, animal lacks oxygen and may choke to death.

Grass tetany

It is a disorder associated with Mg deficiency in the blood of animals especially ruminants. High nitrate content of forage and a low energy intake will also increase the incidence of grass tetany. Introductioin of legumes, which are rich in Mg content in pasture reduce the incidence of this disorder.

Fescue foot

It is a soreness of foot and not a foot rot and occurs in animals grazing tall fescue over a long period of time. There is loss of body weight, rapid breathing, increased temperature and rough haircoat. Later it results in development of slight limp in hind quarters. It is possibly because of alkaloid perloline.

Staggers

Rye grass staggers in sheep and phalaris staggers in sheep, dairy and beef animals are reported to be due to alkaloids present in rye-grass and phalaris.

Nitrate-nitrite poisoning

Results when cattle consume too much forage with high nitrate (in excess of 0.35% to 0.45%) in diet (Walton, 1983). Sudan grass and pearlmillet contain high nitrate, added nitrogen and drought. Nitrite absorbed into the blood stream changes haemoglobin into methaemoglobin which is unable to transport oxygen. Death occurs from oxygen starvation of body tissues. Plants with nitrate content of 1.5% or more potassium nitrate (KNO_3) in the dry matter are potentially hazardous to ruminant livestock. Nitrate poisoning may kill livestock and cause abortion of fetus in pregnant animal.

Prussic acid or cyanide poisoning

Prussic acid- HCN or hydrocyanic acid is problem in sorghums, Johnson grass and Sudan grass. It is high in these forages following drought, frost or freeze. High application of nitrogen also increases the concentration of glucosides. HCN released from plants in the rumen of stock is rapidly absorbed into the blood stream. In large amounts, this blocks the use of oxygen by cells of body, leading to the death of the animals.

Bleeding disease

During heating or spoilage of hay of sweet clover, coumarin gets converted into dicoumaral which prevents blood clotting. Animals with slight wounds or internal hemorrhaging may bleed to death if fed on spoilt sweet clover hay.

Oxalate

High oxalate content in Hybrid Napier, Setaria may cause hypocalcaemia. Formation of calcium oxalate ties up blood calcium and this causes poor coagulation of the blood. Oxalic acid content in *Paspalum plicatulum* is low (0.027%) while in *Pennisetum purpureum* it is high (2.6%). Soluble oxalates are rapidly absorbed into blood stream, where they bind to calcium ions rapidly reducing the amount of calcium available for use by the body tissues. Calcium oxalate crystals are formed in the kidney tubules, interferring with kidney function. Similar crystal formed blood vessels may damage the rumen wall. Plants containing more than 2.0% soluble oxalates have the potential to cause acute oxalate poisoning in very hungry stock.

Scouring

Accumulation of excessive amount of molybdenum in forages in localized area causes scouring.

Alkali Disease

Excessive selenium causes alkali disease.

Mottling and wear of the teeth

It is due to excess level of fluoride in herbage.

Ergotism

It is caused by feeding of ergot (*Claviceps* spp) infected forage which produces alkaloids that result in abortion and sloughing of hooves.

Mimosine

It is a toxic amino acid found in leucaena. Sole feeding results in wool shedding and abortion in sheep and loss of hair on the rum and tail and loss of weight in cattle. Prolonged feeding results in enlargement of thyroid glands.

Indospicine

It is a hepatotoxic amino acid found in *Indigofera*. It causes liver degeneration and abortion in cattle and sheep and staggers and even death in horses, due to interference with metabolism.

9

REFERENCES

Ahlgren, G.H. 1956. *Forage Crops*. Mc-Graw Hill Co., New York.

Aiyer, A.K.Y.N. 1962. *Feeds & fodders*. Bangalore Printing & Publ. Co. Ltd. Bangalore.

Alexander, C.W., Sullivan, J.T. and Mc Cloud D.E. 1962. A method for estimating forage yield. *Agron. J*. 54: 468-469.

Anonymous, 1961. *Research Techniques in Use at the Grassland Research Institute Hurley.* Bulletin No 45, CAB Fernham Royal, Bucks, England.

Anonymous, 1970. *Report of the Grazing Advisory Committee on the Grazing Policy in Himachal Pradesh*. Govt. of Himachal Pradesh pp. 50.

Anonymous, 1979. *Tropical Legumes : Resource for the Future*. National Academy of Science, Washington, D.C.

Anonymous, 1980. *Livestock Units Handbook.* Ministry of Agriculture, Fishery and Food. Northumberland, U.K.

Anonymous, 1982. *Annual Report*. Welsh Plant Breeding Station, Aberystwyth, U.K. pp. 256

Anonymous, 1984. *Annual Report.* The Grassland Research Institute, Hurley, U.K.

Anonymous, 1987. *Handbook of Agriculture*. Indian Council of Agriculture Research, New Delhi.

Anonymous, 1995. *Indian Agriculture in Brief.* Ministry of Agriculture. GOI, New Delhi.

Archer, S.G. and Bunch, C.E. 1953. *The American Grass Book*. University of Oklahoma Press, Norman U.S.A.

Arora, R.K. and Chandel, K.P.S. 1972. Botanical source of wild herbage legumes in India. *Tropical Grasslands*. 6:213-221.

Barnard, C. 1964. *Grasses and Grasslands*. Mc-Millan & Co. Ltd. London.

Beale, I.F. 1973. Tree density effects on yield of herbage and tree components in south west Queenland mulga (*Acacia aneura*). *Tropical Grassland*. 7: 1-35

Bhimaya, C.P. and Ahuja, L.D. 1969. Criteria for the determining condition class of rangelands in Western Rajasthan. *Ann. Arid Zone*. 8: 73-79.

Bogdan, A.V. 1977. *Tropical Pastures and Fodder Plants*. Longman, London.

Brockman, J.S. 1985. Weeds, Pests and Diseases of Grassland and Herbage Legumes. British Crop Protection Council, Cambridge, U.K.

Brown, D. 1954. *Methods of Surveying and Measuring Vegetation*. Bull. 42. Common. Bur. Past. Fld. Crops. Hurley , U.K.

Camlin, M.S. and Stewart, R.H. 1976. The assessment of persistence and its application to the evaluation of early perennial rye grasses cultivars. *J. Br. Grassld. Soc*., 31: 1-6.

Canfield, R.H. 1941. Application of the line interception method in sampling range vegetation. *J. For.* 39: 388-394.

Cherry, M.E. 1968. *Grass Conservation Hand Book*. Shiffe Booke Ltd, London

Chopra, R.N, Bhadŵar, R.L. and Ghosh, S. 1965. *Poisonous Plants of India*. I.C.A.R. New Delhi.

Chopra, R.N., Handa, K.L. Kapoor,I.D. and Singh J. 1956. Nutritive value of grasses of Jammu & Kashmir. *Indian J.Agric. Sci*. XXXIV : 415-457

Christie, E.K. 1984. Natural Grasslands. In. *Control of Crop Productivity*. Ed. C.J. Pearson. Academic Press. New York.

Cooper, J.P. 1970. Potential production and energy conversion in temperate and tropical grasses. *Herb. Abstr*. 40: 1-15.

Cooper, M.M. 1967. *Grass Farming*. Lyods Chambers, Ipsaich

Crampton, E.W. and Harris, L.E. 1969. *Applied Animal Nutrition*. Freeman & Co., San Fransisco.

Crowder, L.V. and Chheda, H.R. 1985. *Tropical Grassland Husbandry*. Longman Group Ltd., London.

Dabadghao, P.M. and Shankarnarayan, K.A. 1973. *The Grass Cover of India*, ICAR, New Delhi.

Davis, W and Skidmore, C.L. 1966. *Tropical Pasture,* Faber and Faber Ltd., London.

Dogra, K.K., Bhandari, J.C. and Singh, L.N. 1979. Dry matter contents and chemical composition of certain indigenous grasses of *Themeda - Arundinella* cover in Himachal Pradesh. *Forage Res*. 5: 49-56.

Dogra, K.K., Katoch, B.S. and Singh, B. 1994. Botanical composition, nutritional status and carrying capacity of native alpine pasture of Bara Bhangal for sheep. *Indian J. Anim. Sci*. 64: 489-492.

Donahue, R.L., Evans, E.F. and Jones, L.I. 1956. *The Range and Pasture Book*. Prentice Hall, New Jersey, U.S.A.

Donaldson, E. and Edwards, R.A. 1983. *Silage Additives*. Technical Note, 319A ESCA, Edinburgh.

Dowling, R.M. and Makenzie, R.A. 1993. Poisonous Plants. D.P.I., Brisbane, Australia.

Driven, J.G.P. 1977. Beef and milk production from cultivated tropical pastures. A comparison with temperate pastures. *Stikstof*, 20:2-15.

Dyksterhuis, E.I. 1949. Condition and management of range land based on quantitative ecology. *J.Range Mgt*. 2:104-115.

Evans, R.A. and Love, M.R. 1957. The step point method of sampling, a practical tool in range research. *J. Range Mgmt*. 10: 208-212.

Evans, L.T., Wardlaw, I.F. and Williams C.N. 1964. Environmental control of growth. In. *Grasses and Grasslands*. Ed. Barnard, C. Macmilan, London.

F.A.O. 1979. *Production Year Book*. Vol.32. FAO, Rome.

Frame, J. 1970. Fundamental of grassland management Part 10: The grazing animal. *Scottish Agne*. 50: 28-44.

Gohl, B.O. 1975. *Tropical feeds. Feeds information, summaries and Nutritive value*. FAO, Rome.

Gupta, R.K. 1978. *Vegetation* 37(2): 117-118.

Hagger, R.J. 1970. Seasonal production of *Andropogon gayanus* I. Seasonal changes in yield components and chemical composition *J. Agric. Sci* (Camb.)74:487-94.

Heady, H.F. 1975. *Rangeland Management*. Mc Graw Hill New York.

Heath, M E, Barnes, R.F. and Metcalfe, D.S. 1985. *Forages. The Science of Grassland Agriculture*. Iowa State University Press, Iowa, USA.

Heath, M.E., Metcalfe, D.S. and Barnes, R.E. 1973. *Forages. The Science of Grassland Agriculture* Iowa State University Press. Ames, Iowa, U.S.A.

Henzell, E.F. 1983. Contribution of forages to world wide food production. XIV *Intern. Grassland Congress*. pp.42-47. Boulder, Colorado, USA.

Henzell, E.F. and Mannetje, L.t. 1980. Grassland and forage research in tropical and subtropical climates. In "*Perspectives in world Agriculture*" commonwealth Agricultural Bureau, London.

Hitchcock, A.S. 1950. *Manual of the Grasses of the United States*. U.S.D.A., USA.

Hodgson, R.E., Davis, R.E., Hosterman W.H. and Hienton, T.E. 1948. Principles of making hay. In *Yearbook of Agriculture*, U.S.D.A., Washington B.C. pp. 161-167.

Humphreys, L.R. 1969. *A Guide to Better Pastures for the Tropics and Sub-tropics*. Wright, Stephenson and Company Ltd. Melbourne, Australia, pp.79.

Humphreys, L.R. 1978. *Tropical Pastures and Fodders Crops*. Intern. Tropical Agric. Series. Longman Group Ltd. London.

Humphreys, L.R. 1991. *Tropical Pasture Utilization*. Cambridge University Press Cambridge, U.K.

Humphreys, R.R. 1945. Some fundamentals of the classification of range condition. *J. Forestry*. 43: 646-647.

Hutton, E.M. 1970. Tropical pastures. *Adv. Agron*. 22:1-74

Ivins, J.D. 1959. *The Measurement of Grassland Productivity*. Butterworths Publ. London.

James, B.J.F. 1969. *Intensive Utilization of Pasture*. Angus & Robertson. London.

Jones, R.J. and Sandland, R.L. 1974. The relationship between animal gain and stocking rate. *J. of Agric. Sci.* 83:335-342.

Joshi, S.P., Raizaida, A. and Srivastava, M.M. 1990. Carrying capacity of an alpine grazing land of Garhwal Himalayas. *Range Management of Agro Forestry* 11(2):141-150.

Katoch, D.C., Singh, L.N. and Modgal, S.C. 1975. A note on fertilizer use in grasslands of midhills and its economics. *Himachal J. Agric. Res.*3:53-55.

Kaul, R.N. and Gogte, M.G. 1983. *Greening of Forest Grasslands in Nasik District*. National Afforestation and Eco. Development Board, GOI, New Delhi.

Kemp, D.R. 1984. Temperate Pastures. In *Control of Crop Productivity* Ed. C.J. Pearson. Acadmic Press, New York.

Kumar, P. and Sood, B.R. 1998. Renovation of *Imperata cylindrica* dominant natural grassland through introduction of improved grass species. *Indian J. Agron*. 43: 183-187.

Lal, R.B. 1990, *Principles and Practices of Range Management.* International Book Distributors, Dehradun.

Lancaster, R.R. James. E., Bailey, R.V. and Harris, R.R. 1949. *Pastures*. Turner E. Smith & Co. Atlanta, Georgia.

Lander, P.E. 1949. *The Feeding of Farm Animals in India*. Mac Millan & Co. Ltd. P. 492.

Langer, R.H.M. 1973. *Pasture and Pasture Plants*. A.H. & A.W. Reed Publication, Wellington.

Lauenroth, W.K. 1979. *Grassland Primary Production. North American Grasslands in Perspective.* In "Perspectives in Grassland Ecology, Ecological Studies" Vol.32 (N.R. French, ed.) pp.3-24 Springer-Verlag, New York.

Lazenby, A and Swain, F.G. 1972. *Intensive Pasture Production*. Angus & Robertson Publ., London.

Leith, H. 1975. Modelling the primary productivity of the world. In : Primary Productivity of the Biosphere Ecological studies Vol. 14 (H. Leith & R.H. Whittaker eds.) pp. 185-202. Springer Verlag, New York.

Leng, R.A. 1987. Determining the nutritive value of herbages. In *Forages in South East Asian and South Pacific Agriculture*. Australian Centre for Intern. Agric. Res., Canberra.

Lush, R.H. 1952. *Pasture Production and Management*. Blackiston Co., Inc. New orkY., pp.193.

Mannetje, L.'t 1978. *Measurement of Grassland Vegetation and Animal Production*. C.A.B. Bucks, England.

Mannetje, L.'t and Haydock, 1963. The dry weight rank method for botanical analysis of pasture. *J. British Grassld. Soc*. 18 : 268-275.

Mays, D.A. 1974. *Forage Utilization*, American Soc. Agron. Madison, U.S.A.

McDowell, R.E. 1985. Meeting constraints to livestock production system in Asia. Proc. of Regional Workshop on Livestock Production Management, pp. 37-66, Asia Dev. Bank Manila.

McLlroy, R.J. 1972. *An Introduction to Tropical Grassland Husbandry*. Oxford Univ. Press. U.K.

Melkania, N.P. and Singh, J.S. 1989. *Perspective in Ecology* (Eds. J.S. Singh and B. Gopal). Jagmander Book Agency, New Delhi. pp.67-103.

Mengel, K. and Kirby, E.A. 1979. *Principles of Plant Nutrition*. Intern. Potash Inst., Berne.

Millar, C.E., Turk, L.M. and Folh, H.D. 1958. *Fundamentals of Soil Science*. John Wiley and Sons. New York.

Miller, D.A. 1984. *Forage Crops*. Mc Graw Hill Co. New York. pp. 530

Milner, C. and Hughes H.E. 1968. *Methods for the Measurement of the Primary Production of Grasslands*. Blackwell Oxford.

Moore, H.I. 1950. Grassland Husbandry, George Allen and Unwin Ltd., London.

Mott, G.O. 1960. Grazing pressure and the measurement of pasture production. *Proc. 8th Intern. Grassland. Cong*. pp. 606-11. Aldan Press. Oxford.

Mott, G.O. 1962. *Pasture and Range Research Techniques*. Comstock Publishing Associate Ithaca, New York.

Mudgal, V.D. and Pradhan K. 1988. *Proc. of construction on non-conventional feed resources and fibres agricultural residues*. IDRC. ICAR. H.A.U., Hisar pp 139-145.

Narayanan, T.R. and Dabadghao, P.M. 1972. *Forage Crops of India*. India Council of Agricultural Research, New Delhi.

O'Reilly, M.V. 1975. *Better Pastures for the tropics*. Arthur Yates and Co Pvt. Ltd. Revesby, Austrailia.

Ostrowski, H. 1972. *White clover in Queensland*. Advisory leaflet No. 1140. Division of Plant Industry Department of Plant Industries, Australia.

Pal, R.N., Dogra, K.K., Singh, L.N. and Negi, S.S. 1979. Chemical composition of some free fodders in Himachal Pradesh. *Forage Res*. 5: 109-116.

Pandey, A.N. 1981. Seasonal behaviour of grazing animals and an assessment of carrying capacity of the grazing lands. *Trop. Ecol*. 22: 110-186.

Parker, K.W. and Harris, R.W. 1959. The three step method for measuring condition and trend of forest ranges: a resume of its history development and use. In. *Techniques and Methods of Measuring Understory Vegetation*. U.S. Forest Service. South and South East Forest Exp. Sta. Proc. pp.55-69.

Paroda, R.S. and Mal, B. 1995. Research practices for improvement of rangelands. *Annals of Arid Zone*: 34: 227-231.

Pearson, C.J. and Ison, R.L. 1987. *Agronomy of Grassland Systems*. Cambridge Univ. Press Cambridge.

Peoples, M.B. and Craswell, E.T. 1992. Biological nitrogen fixation: Investment, expectations and actual contributions to agriculture. *Plant and Soil*. 14: 13-39.

Phillimore, P. 1984. In: *Changing South Asia: Economy and Society*. (Eds.) K. Ballahatchet and D. Taylor. Asian Research Service, Hong Kong. pp.123-133.

Pimental, D.W., Dritselilo, W., Krunmel, J. and Kurtzman, T. 1976. Energy and land constraints in food protein production. *Science* 190:754-761.

Raymond, W.F. 1969. The nutritive value of forage crops. *Adv. Agron.*, 21:1-108.

Rehm, S. and Espig., G. 1991. *The cultivated Plants of the Tropics and Subtropics*. Scientific Books.

Sampson, A.W. 1952. *Range Management Principle and Practices*. Chapman and Hill, London.

Sardar, M.R. 1989. *Rangelands of Pakistan: Their Potential for Development*. Paper presented at FAO Regional Workshop on Himalayan Pasture lands. Peshawar. pp. 19.

Sehgal, J.L., Mandal, C and Vadiye, S. 1992. *Agro Ecological Regions of India*. National Bureau of Soil Survey and Land use Planning, Nagpur.

Semple, A.T. 1952. *Improving the Worlds Grassland*. Leonard Hill Ltd. London.

Semple, A.T. 1970. *Grassland Improvement*. Leonard Hill Book, London.

Sen, K.C. and Ray, S.N. 1964. *Nutritive Values of Indian Cattle Feeds and the Feeding of Animals*. I.C.A.R. Bull, No. 25, New Delhi.

Serviss, G.H. and Ahlgren, G.H. 1955. *Grassland Farming*. John Wiley & Co., New York.

Shantz, H.L. 1954. The place of grasslands in the earth's cover of vegetation. *Ecology*. 35: 143-151.

Sharma, B.R. and Koranne, K.D. 1988. Management of primary productivity on the degraded rangeland of North-Western Himalayas. *Third Intern. Rangeland Cong*. 1: 67-69.

Shaw, N.H. and Bryn, W.W. 1976. *Tropical Pasture Research. Principle and Methods*. C.B.P & F.C. Hurlely, U.K.

Shultis, A. 1962. *Estimating Rangeland Values and Routs*. Univ. of California Agric. Ext. Serv. Publ. AXT.70

Singh, L.N. 1990. Performance of Leucaena under different spacings and cuttings management in sub-montane region. *Range. Mgmt. & Agroforestry*. 11: 213-216.

Singh, L.N. 1990. Seasonal variation in dry matter production of certain tropical grasses in sub-montane region of Western Himalayas. *Range Mgmt. & Agroforestry*. 11: 29-32.

Singh, L.N. 1992. Grazing routes in Himachal Pradesh. *Range Mgt. and Agroforestry*. 13(2):147-152.

Singh, L.N. and Katoch, D.C. 1975. Forage yield potential of some grasses. *Indian J.Agric.Res.* 9:25-29.

Singh, L.N. and Sharma, B.K. 1993. Establishment, persistence and forage yields of certain grasses in Kullu valley. *Range Mgmt. & Agroforestry*. 14(1):11-16.

Singh, L.N., Dogra, K.K. and Katoch, D.C. 1975. Botanical composition herbage production and nutrient status of grasslands of Himachal Pradesh. *Forage Res*. 1: 40-46.

Singh, L.N., Bhandari, J. C. and Dogra, K.K., 1977. Preliminary evaluation of some temperate pasture legumes and grasses in mid hills of Himachal Pradesh. *Forage Res*. 3: 27-33.

Singh, L.N., Katoch, D.C., and Dogra, K.K. 1980. Effect of legume introduction on forage yield and quality of natural grassland. *Forage Res*. 7: 71-76.

Singh, L.N., Katoch, D.C. and Dogra, K.K. 1981. Performance of certain tropical and subtropical grasses in mid-hills of Himachal Pradesh. *Forage Res.* 7:143-148.

Singh, L.N., Singh, A. and Singh, J. 1998. *Forage Grasses and Legumes.* Scientific Publishers, Jodhpur.

Singh, P. 1988. *Indian Rangelands Status and Improvement*. Plenary address. Third Intern.Rangeld. Cong., New Delhi.

Singh, P. 1995. Rangelands and their improvement in India. *Annals of Arid Zone* 34(3): 157-161.

Singh, P. and Majumdar, A.B. 1992. Current status of field and forage in management of livestock in India. *Agric. Situation in India.* XLVII(5):375-382.

Singh, R.P. 1986. *Handbook of Temperate Forage Production*. International Book Distributors, Dehradun.

Skerman, P.J., Cameron, D.G. and Riveros, F. 1988. *Tropical Forage Legumes*. FAO, Rome, pp.692.

Skerman, P.J. and Riveros, F. 1990. *Tropical Grasses*. Food and Agriculture Organisation of the United Nations, Rome pp.832.

Sood, B.R. and Kumar, P. 1995. Production and economics of natural grassland improvement. *Range Mgmt. & Agroforestry*. 16:17-22.

Spedding, C.R.W. 1971. *Grassland Ecology*. Clarandon Press, Oxford, U.K.

Steele, J.G. 1951. *The Measure of Our Land*. Soil Cons. Serv., Govt. Print. off. Washington D.C.

Stefferud, A. 1948. *Grass*. The year Book of Agriculture. U.S.D.A. Washington.

Stewart, G. and Hutchings, S.S. 1936. The point observation - plot square foot density method of vegetation survey. *J. Am. Soc. Agron*. 28: 714-726.

Stobbs, T.A. and Thompson, P.A.C. 1975. Milk production from Tropical Pastures. *World Animal Review*. F.A.O. No.13 pp.27- 31.

Stoddart, L.A., Smith, A.D. and Box T.W. 1965. *Range Management*. Mc-Graw-Hills Book Co., London.

Sucking, F.E.T. 1965. *Hill Pasture Improvement* D.S.I.R. New Zealand.

Sundriyal, R.C. 1989. Assessment of the grazing ability of an alpine pasture in the Garhwal Himalaya. *Environments and Ecology*. 7: 247-249.

Sundriyal, R.C. 1995. Grassland forage production and management in the Himalaya : A review *J. Hill Res*. 8(2): 135-150.

Sundriyal, R.C; Sharma E. and Negi, S.S. 1993. Effect of cutting height and frequency on the above ground biomass in a central Himalayan grassland in India. *Tropical Grasslands*. 27:37- 42.

Thomas, C. and Young, J.W.O. 1984. *Milk from Grass*. I.C.I.Billingham, U.K.

Tilak, K.V.B.R. and Singh, G. 1996. Integrated nutrient management in sustainable agariculture. *Fertilizer news*. 41(4): 29- 35.

Tilley, M.A. and Terry, R.A. 1963. A two stage technique for the *in vitro* digeston of forage crops. *J.Br.Grassland Soc*. 18: p.104.

Troll, C. 1996. *Season climates of the earth. The seasonal course of natural phenomena in the different climatic zones of the earth.* In Rodenwalt, E. and Juratz, H.J., Eds World Maps of Climatology. Springer Verlag. Berlin.

Trewartha, G.T. 1954. *An Introduction to Climate*. McGraw Hills, New York.

Tucker, R.P. 1986. The evolution of transhumant grazing in the Punjab-Himalayan. *Mountain Res. and Development* 6:17-28.

Tueller, P.T. 1995. Remote sensing in the management of rangelands. *Annals of Arid Zone*. 34(3): 191-207.

Vallentine, J.F. 1989. *Range Development and Improvement*. Academic Press Inc. San Diego, U.S.A.

Van Soest, P.J. 1966. *Non-nutritive residues. A system analysis for the replacement of crude fiber*. J. A.O.A.C. 49:546-62.

Walker, T.W. 1956. *J. Sci. Fd. Agric*. 7: 66-71.

Walters, R.J.K. 1980. *The systematic evaluation of grass varieties at the Welsh Plant Breeding Station*. Annual Report, Aberystwyth. U.K.

Walton, P.D. 1983. *Production and Management of Cultivated Forages*. Reston Pub. Co. Virginia U.S.A.

Watson, S.J. and Nash, M.J. 1960. *Conservation of Grass and Forage Crops*. Oliver & Boyd, London.

Wheeler, W.A. 1950. *Forage & Pasture Crops*. D.Van. Nostrand Co. Inc. New York.

Wheeler, W.A. and Hill, D.D. 1957. *Grassland Seeds*. D.Van Nostrand Co. Inc. New York.

White, J.G.H. 1973. Improvement of hill country pasture. In *Pasture and Pasture Plants*. Ed. R.H.M. Langer. A.H. & A.W. Reed Pub. Wellington.

Whiteman, P.C. 1980. *Tropical Pasture Science* Oxford University Press, New York.

Whyte, R.O. 1964. *The Grassland and Fodder Resources of India*. Indian Council of Agricultural Research, New Delhi.

Wilkins, R. and Down, K. 1984. Realizing the potential of British Grassland. *Grassland Research Today*. 2-3. A.F.R.C. London.

Wilkinson, S.R. and Lowrey, R.W. 1973. Cycling of mineral nutrients in pasture ecosystem. In: *Chemistry and Bio- chemistry of Herbage* Eds. G.W. Buler & R.W. Bailey. pp. 247- 345 London, Acadmic Press.

Williams, R.D. 1984. *Crop Protection Hand Book - Grass and Clover Swards*. B.C.P.C. Publication, England.

Wilson, J.R. 1984. Tropical pasture. In: *Control of Crop Productivity*. Ed. C.J. Pearson, Academic Press, New York.

Wilson, J.R. and Minson, D.J. 1980. Prospects of improving the digestibility and intake of tropical grasses. *Tropical Grasslands,* 14 : 253-259.

Yadava, P.S. and Singh, J.S. 1977. *Grassland Vegetation : Its Structure, Function, Utilization and Management*. Today and Tomorrows Printers and Publishers, New Delhi, p 132.

GLOSSARY

Alkaloid	:	An organic chemical compound containing nitrogen usually with a bitter taste and having strong physiological activity.
Alpine	:	Areas on mountain heights above timber line.
Animal days per unit area of land	:	The number of animals per unit area multiplied by the number of days that animals are on pasture. This is considered a measure of quantity i.e. herbage yield of pasture.
Animal unit	:	A standardized term used to describe the number of animals on pasture. One animal unit is considered a 2 year old beef cow, bull, stear or heifer.
Animal unit month (AUM)	:	The potential forage intake of one animal unit for a period of one month (30 days), based generally on 750 lb of dry matter consumed.
Ascending	:	Stems growing obliquely at first but finally upwards
Available water capacity	:	Available water capacity is index of water holding capacity of soil.
Available soil water	:	The amount of water held between wilting point and field capacity.
Biomass	:	Weight of living plant material.
Browse	:	Leaf and twig growth of shrubs, woody vines and trees available and acceptable for animal consumption.

Brush : Shrubs or small trees considered undesirable from the stand point of planned use of area.

C3 plants : Plants that convert carbondioxide into three carbon compounds during photosynthesis.

C4 plants : Plants that convert carbondioxide into four carbon compounds during photosynthesis.

Calorie (small) : The amount of energy as heat required to raise the temperature of one gram of water by $1^{o}C$.

Canopy : The sward growth as it intercepts or absorbs light.

Carbon balance : Balance between photosynthesis and respiration which leads usually to a daily net gain in dry matter.

Carrying capacity : Number of animals that can graze in an unit area without over or under grazing in an average season.

Carrying over feed : Pasture accumulated during periods favourable for growth is "saved" for periods of deficiency.

Chaparral : Dense stands of shrub by plants dominated by broad leaf and narrow leaf non deciduous species.

Closed canopy : A canopy which appears to have achieved complete cover or intercepts 95% of visible light.

Continuous grazing : The system of grazing in which whole grassland is opened for grazing at the same time.

Critical area stocking rate : Stocking rate below which pasture stability and a satisfactory legume content is maintained.

Crown	:	In clovers and other hemicryptophytes, the top of the tap root bearing buds from which the basal leaf rosette and shoots arise. Not appropriate to grasses.
Crude protein	:	Total ammonical nitrogen x 6.25 based on the fact that feed protein on the average contains 16 % nitrogen.
Culm	:	The extended stem of a grass tiller bearing the inflorescence.
Cultivar	:	Cultivated variety
D-value	:	It is the percentage of digestibility of organic matter in dry matter of fodder.
Decumbent	:	Of stems lying on the ground but with ends pointing upwards.
Deferred grazing	:	Grazing in which grazing is deferred in one portion till the seed forms in grasses.
Denitrification	:	The reduction of nitrate to dinitrogen gas (N_2) and nitrous oxide (N_2O)
Dry matter	:	That part of feed which is not water. This is the sum of the crude protein, crude fat, crude fibre, nitrogen free extract and ash.
Feed	:	Any non injurious edible material having nutritive value for animals.
Feed year planning	:	To match the supply of feed with animal demand for the whole year and is carried out on a whole farm basis for specific livestock units or for particular forage sub systems.
Foggage	:	The coarse grass not mowed or eaten in summer or autumn.
Foliage	:	A collective term for the leaves of a plant or community.
Forage	:	Plant material, including herbage but excluding concentrates, used as feed for domestic herbivores.

Forage density	:	This refers to only that vegetation which is within reach of livestock. Vegetation not available to stock is excluded from the estimates of forage density.
Forage acre	:	It is theoretical acre totally covered with vegetation all of which is properly usable by livestock.
Grassland	:	Plant community, natural or sown, dominated by grasses, herbaceous legumes and other herbaceous species; a small proportion of trees or shrubs may be present.
Grazing capacity	:	Total number of animal unit months (AUM) produced and available for grazing from a pasture unit, a grazing allotment, the total ranch or other specified land area.
Grazing management	:	The manipulation of grazing animals to accomplish desired results in terms of animal, plant, land or economic responses.
Grazing pressure	:	A relationship between demand for forage by animals and a combination of daily herbage increment and standing crop of vegetation.
Ground cover	:	The proportion of the ground area covered by the canopy when viewed vertically.
Hay	:	Forage dried without deterioration of dry matter and nutrients, having its natural green colour and palatability and capable of being stored over a long period of time.
Haylage	:	Low moisture silage (35 to 55 per cent moisture). Grass and legume crops are cut and wilted in the field to a lower moisture level than normal for grass silage but the crop is not sufficiently dry for baleing.
Herbage	:	The above ground parts of a sward viewed as an accumulation of plant material with

		characteristics of mass and nutritive value.
Herbage accumulation	:	The actual measured increase of herbage per unit area i.e. herbage growth minus herbage lost.
Herbage growth	:	Increase in weight of herbage per unit area over a given time interval due to its production of new material.
Herbage residual	:	The herbage remaining after grazing.
Inflorescence emergence	:	Mean date of heading date or ear emergence. The mean date on which the third inflorescence of each plant first emerges (in spaced plant); for a sward the date on which 50 % of the inflorescences in fertile tillers have emerged.
Juvenility	:	Early phase of growth from seed during which flowering can not be induced by any treatment.
MD value	:	The concentration of metabolizable energy (ME) per kg of pasture dry matter.
Meadow	:	A hay field, or a pasture not grazed, but one in which plants are allowed to grow for the production of hay or seed or both.
Natural pasture or range	:	Land on which the plant community is composed primarily of native grass and shrubs valuable for forage and in sufficient quantity to justify grazing use.
Net primary productivity	:	Net change in weight of grassland plants between any two points in time, usually over a year.
Nitrification	:	Oxidation of ammonium (NH_4) to nitrate (NO_3) by bacteria *Nitrosomonas* and *Nitrobacter*.
Nitrogen free extract	:	That part of feed dry matter which is not crude protein, crude fat, crude fibre, or ash. It consists mostly of sugars and starches.

Noxious plant : A plant which is undesirable in light of planned land use or which is unwhole some to range lands or range animals.

Nutrition index : Quantification of the extent to which growth is constrained by nutrition. It ranges from zero when there is no growth to 1 when growth is not restricted by nutrition.

Overdrilling : Drilling of seed and fertilizer into an existing crop stubble or pasture without previous cultivation.

Oversowing : Surface broadcasting of seed (and fertilizer) into an existing pasture, although it is occasionally extended to mean broadcasting into an existing crop or crop stubble.

Palatability : Summation of plant characteristics that determine the relish with which a particular plant species or plant part is consumed by an animal.

Pampas : Vast grassy mid latitude or temperate treeless plains of South America

Pasturage : Plants materials harvested directly by grazing animals.

Pasture : Land occupied by perennial forage or by self seeding annuals, usually both which are used as food by grazing animals

Pelleting : Coating of grass and legume seed with a layer of fine powder held on to the seed by a sticker.

Periodic grazing : System of grazing in which whole grassland is grazed or cut for hay for a year or two but closed for self seeding and recovery in the following year.

Pitting : Formation of small basins or pits or other indentations of soil surface for enhancing water penetration

Plant density	:	The number of plants or tillers per unit of area.
Poaching	:	Trampling of grasslands with soft or wet soil causing packing of soil particles and loss of larger pores in the soil mass.
Poisonous plant	:	A plant which injures animal health when taken internally because of toxic agent present.
Prairie	:	An extensive formation dominated by grasses and largely composed of bunch grasses found in North America.
Procumbent	:	Stems, trailing loosely along the ground.
Produce per unit area	:	A measure of liveweight gain or milk production i.e. product per hactare = produce per animal per day x animal days per hectare.
Prostrate	:	Plants lying flat on the ground.
Pure live seed	:	Term indicative of quality and quantity of seed based on percentage germination and percentage purity.
Rangeland	:	The rangelands are the uncultivated natural forest lands and grasslands which provide browsing and grazing grounds to the livestock and wild animals.
Reclamation	:	Putting a natural resource to a new or altered use.
Rehabilitation	:	Improvements to a natural resource that return it to a good condition but not the condition prior to disturbance.
Rhizome	:	An underground stem, often creeping and branched, sending off rootlets, the apex progressively sending up stems or leaves.
Ripping	:	Subsoiling or deep chiseling to break compacted layers in the soil that inhibit root development and moisture penetration.

Rotational grazing	:	Grazing system in which grazing is allowed in all sub units of grassland area alternately and uniformely.
Roughage	:	The term applies to hay, fodder, stover, pasture plants, silage and roots used for animal feeding purposes.
Savanna	:	A track of level land, covered with low vegetation, treeless or dotted with trees or patches of wood in vast area of Africa, South America and Northern Australia.
Shrub	:	Woody plant, usually having several erect, spreading or prostrate stems and generally under 15 feet (commonly under 10 feet) at maturity.
Silage	:	Moist succulent feed produced as a result of controlled fermentation of fresh forage when stored in a silo under anaerobic conditions.
Silo	:	A storehouse of silage.
Sod	:	A piece of turf dug up, or pulled up by grazing animals.
Sod-seeding	:	Refers to overdrilling of seed into an existing pasture.
Soiling crops	:	Crops cut green and fed green from the field.
Steppe	:	One of the large tracts in South Eastern Europe or Asia which are generally level, grassy and without trees.
Stipule	:	One of a pair of appendages at the base of a leaf of some plants.
Stocking rate	:	Stocking rate is the actual number of animals or animal units on a specific area for a specific period of time, usually for a grazing season.
Stover	:	The term applies to the corn stalk and leaves from which the ears have been removed.

Stubble	:	The part of sward canopy left behind after defoliation
Succession	:	The sequence of communities that develop in an area from the initial stages of colonization until a stable mature climax community is achieved.
Sward	:	An area of grassland with a short continuous foliage cover including both above ground and below ground plant parts but excluding any woody plants.
Sward canopy	:	The above ground parts of sward with connotations of the distribution and arrangement of the various plant parts.
Sward structure	:	A general term for the form distribution, arrangement and relative proportions of plant parts in sward, including canopy structure.
Swath	:	Line of grass or grain cut and thrown together by scythe.
Target yield	:	The point where the response drops to 10 kg dry matter of grass for each kilogram of nitrogen applied.
Temperate climate	:	Climate which is neither very hot nor very cold.
Treading	:	Pressing of grassland by feet of grazing livestock causing crushing of vegetation and soil.
Tropical	:	Either of the two circles of the celestial sphere, on each side of the equator at a distance of 23°27' and parallel to it.
Tundra	:	Terrestrial habitat zone between taiga in south and polar region in north characterized by absence of trees, short growing season and frozen ground during much of the year.

Turf	:	The part of the sward which comprises the shoot system plus the upper most layer (say 10 cm) of roots and soil.
Undersowing	:	Refers to establishment of clover, lucerne, or even a complete grass mixture under a crop, but in a cultivated seedbed. It may be broad cast or drilled.
Veldt or Veld	:	Open country side with grassland having some shurbs and trees found in Southern Africa.
Vernalization	:	Induction of flowering by cold (or heat) treatment of seeds or later developmental stages.
Water use efficiency (WUE)	:	Ability of plant to produce dry matter per unit of water consumed. Species which use more water per unit dry matter increament has lower WUE.
Weed	:	Any plant growing out of place.
Woodland pastures	:	Wooded areas with grass and other grazing plants growing in open spaces and among the trees.
Xerophyte	:	A plant adapted to live under dry or desert conditions.
Zero grazing	:	Feeding of fresh cut herbage to housed livestock.

INDEX

C

D

E

F

G

H